THE HUNT

THE OUTCOME IS NEVER CERTAIN

ALASTAIR FOTHERGILL AND HUW CORDEY
FOREWORD BY DAVID ATTENBOROUGH

BBC
BOOKS

CONTENTS

Foreword 6

Chapter One   **THE HARDEST CHALLENGE** 12

Chapter Two   **HIDE AND SEEK** 50

Chapter Three   **NOWHERE TO HIDE** 84

Chapter Four   **RACE AGAINST TIME** 126

Chapter Five   **IN THE GRIP OF THE SEASONS** 160

Chapter Six   **HUNGER AT SEA** 198

Chapter Seven   **TALES FROM THE HUNT** 236

Index 304

Acknowledgements 310

Picture credits 312

FOREWORD
DAVID ATTENBOROUGH

Filming a hunt out on the open savannahs of Africa is one of the trickiest jobs you are likely to get as a wildlife cameraperson. As always, you will have to ensure that the animal you are following is unaware of your presence. But in this case you will have to conceal yourself from not just one but two – the hunted as well as the hunter. If either spots you during the stalk, you may have lost your chance. You will have to think very carefully about exactly where to place yourself. It may be that you should not be behind the hunter but in front

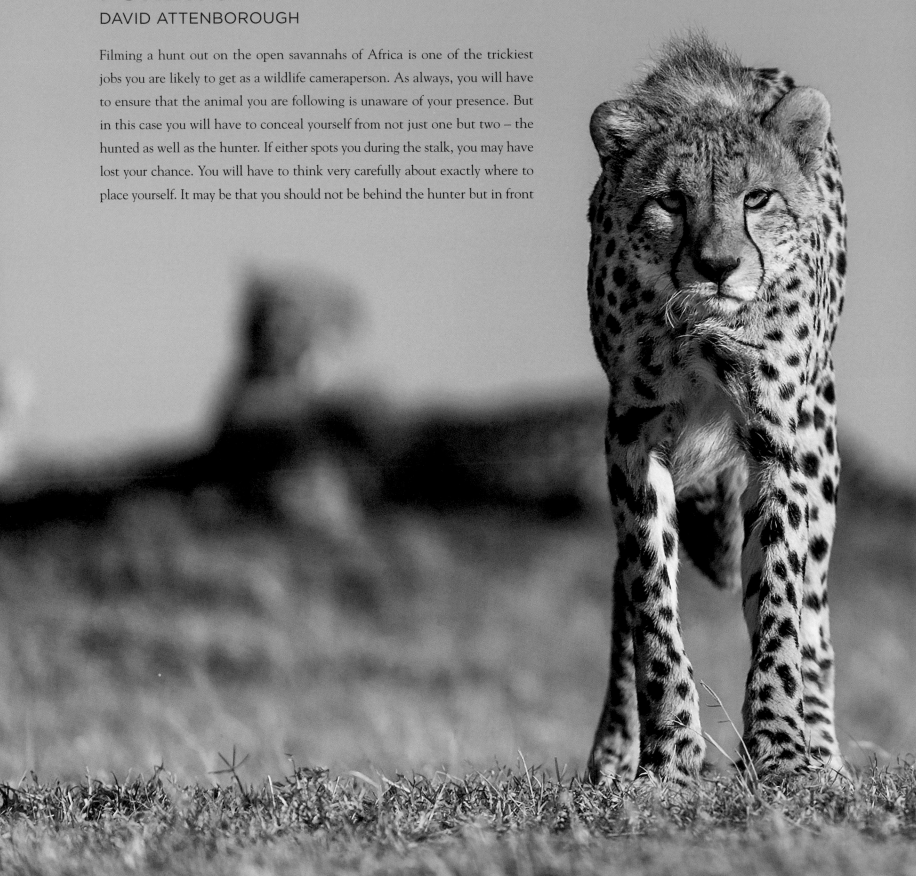

Cheetah with a view. A family of adolescent cheetahs lies up on a raised area in the Serengeti, from where they can keep a watch for lions and hyenas as well as prey. More than 15 per cent of a cheetah's kills will be lost to other predators, though a coalition of brothers may be able to defend a catch.

of it – and so far ahead of the prey that you will be close to the place where the catch will finally be made. To do that, you must know, or be able to judge, the nature of the ground over which the drama will be played out. A bend in a river may act as a barrier that the hunter will exploit; a patch of marshy ground may slow down a fleeting animal; or a slight rise, if you can get there unnoticed, may give you and your camera a crucially better view of the drama that otherwise may be out of sight.

Sixty years ago, when natural history television was in its infancy, it was rumoured that one of the most successful makers of African wildlife programmes had offered five hundred pounds – an almost unbelievable sum in those days –

◀ **Little hunter.** Camerawoman Sophie Darlington gets unexpectedly close to a diminutive hunter – a long-tailed macaque looking for a suitable tool for hammering open shellfish. Its hunting techniques were as skilful and impressive as those of Sophie's other *Hunt* subjects – cheetahs and lions.

▶ **(next page) Return from the hunt.** Two lionesses and their four adolescent cubs return at first light from a hunt. Using the cover of vegetation or darkness to stalk prey, they work as a team to set a trap, one driving the target animal towards another lying in wait. But the chance of a successful hunt is still little more than 15 per cent.

to any cameraperson who could get a clear shot of a lion pouncing on a wildebeest. For many years, there were no takers.

These days, of course, we are better at it. We have much more sophisticated and versatile equipment to help us – powerful long-focus lenses that can give us close-ups from afar; camera mounts that will smooth out shakes and judders so that you can even get acceptable pictures from the back of a jeep as it buckets and crashes through the bush; highly sensitive cameras that can see better than your own eyes on moonlit nights, when so many chases, in fact, take place; and electronic mechanisms that enable you to slow down movements 400 times so that every detail and nuance of the action can be seen clearly and understood.

Today, too, you may have local experts to guide you and scientists who have been studying the behaviour of particular species for decades. If the project is important enough, then you may even have other camerapeople working alongside you to cover different aspects of the dramas that may eventually unfold.

And it is worth all this trouble, for the duels between hunters and hunted are as dramatic as any event in the natural world. These are the moments when an animal's skills are fully stretched – when the issue is life or death.

Duels between hunters and hunted are not, of course, restricted to the open plains of Africa. They are fought out in every ecosystem on the planet and on every scale. Polar bears and harpy eagles are solitary hunters, whereas wild dogs and killer whales operate in teams. Spiders construct elaborate traps of silken filaments, while Sargassum fish disguise themselves with such perfection that they become, in effect, invisible. Indeed, every conceivable skill and accomplishment in the animal kingdom is deployed by some animal somewhere, either to catch – or to escape.

You might think that if the sequences documenting these hunts are to be complete, they should end in a kill. Were they all to do so, however, they would grossly misrepresent nature, for the fact is that the great majority of hunts do not end in death. The subject of this book, and indeed the programmes on which it is based, is not therefore killing. It is the relationship between predators and prey. These duels have, in fact, resulted in the evolution of some of the most refined and sophisticated physical abilities, sensory systems and behavioural tactics to be found in the natural world. The following pages reveal in truly astonishing and unforgettable detail just how true that is.

CHAPTER ONE
THE HARDEST CHALLENGE

THE RELATIONSHIPS BETWEEN PREDATORS AND PREY are the most dramatic in nature – indeed, a matter of life or death. They involve a fascinating range of strategies, whether of capture or escape, crafted according to their environment. Different environments present different challenges. On the short-grass plains of the Serengeti, there is little cover for a stalking leopard. On the Canadian tundra, there is nowhere for a caribou calf to hide from a wolf. In the blue desert of the open ocean, a whale has to travel vast distances in search of food. In the dense tropical forests of Central Africa, predators and prey play hide-and-seek. Most of the time, it's the predators that fail. For any chance of success in countering its prey's defences, a predator needs specific skills tailored to its environment.

▶ **The buffalo hunt.** A charge by bulls sends a lion running for its life. Only a team attack by the pride's lionesses has a chance of bringing down a buffalo.

◀ **(previous page) Ice hunter.** A young polar bear leaps between ice floes. As the ice breaks up in summer, hunting becomes increasingly less successful.

THE NEED FOR SPEED

For many of the planet's top predators, speed is vital to hunting success. The fastest avian predator is the peregrine falcon, which in level flight can easily reach 65–95 kilometres an hour (40–60mph). Few species of waders, ducks and pigeons can outpace it on the level, but the falcon's real weapon is its stoop. Folding its wings tight against its body, it plummets down on its unsuspecting prey at speeds of up to 320kph (200mph). It goes so fast that it would be dangerous to try to grab the bird it's after. Instead it closes its talons and aims to stun it with a punch to the back of the head. The peregrine's strategy is so successful that the species is found on every continent except Antarctica, and possibly a fifth of the world's bird species are on its menu.

Underwater predators can't match their avian counterparts in terms of speed – the viscosity of water slows them down. But the fastest shark, the shortfin mako, which has been recorded at 50kph (31mph), with bursts up to 74kph (46mph), uses its speed for surprise. Like the peregrine, it catches its prey unaware, by swimming below it and then lunging vertically up.

The ocean's fastest predator, though, is the sailfish, which has been recorded at 108kph (68mph). Like all the other billfish, a sailfish is packed with muscle and perfectly streamlined. It combines intense bursts of speed with extraordinary stamina and can keep up a cruising speed of more than 48kph (30mph). Largely solitary, it scours the boundary between cold and warm

◀ **Hit, snatch and grab.** Having used its talons as fists to knock the willet out of the sky and onto the beach, the peregrine makes a pass and grabs it out of the surf. Then it will carry the bird to a plucking perch, where it will break its spine with its beak.

currents for days, searching for smaller fish. Its speed and manoeuvrability really come into play when dealing with a shoal of fish bunched defensively into a baitball; it races in to knock out fish after fish.

▲ **The sprinter.** In mid-air and mid-stride, a cheetah gains on its target in seconds. It has superior speed and agility, but had it started the sprint from too far away, the Thomson's gazelle would have had the stamina to outrun it.

THE FASTEST-RUNNING HUNTER

The top speed that a cheetah can reach is a matter of legend, though the actual speed logged is 93kph (58mph). Everything about it is designed for speed. It has a slight build, a narrow chest, long legs and large heart and lungs. Its muscles are packed with the fast-twitch fibres vital for sprinting. Half of this muscle is packed around the longest and most flexible spine of any large cat – the key to its speed. It allows a higher stride frequency and a stride length of nearly 10 metres (32 feet). For more than half of every stride, the cheetah is completely airborne. But there is a price for this flexible spine: while the cheetah runs, it is constantly moving the heaviest parts of its body – its hips and shoulders – which demands a lot of energy. This means that it can only keep up its record-beating sprint for ten seconds.

A cheetah's whole hunting strategy is shaped around this ten-second limit. Its favourite prey is smaller antelopes such as springbok, impala and

Thomson's gazelle. These have a top speed of about 77kph (48mph) – less than a cheetah's, but only when it comes to the sprint. So a cheetah's hunting success depends as much on its ability to stalk close enough for the sprint as it does to run down its prey. It has to creep unnoticed to within 50 metres (165 feet) before bursting into the high-speed chase. Any farther away and it will tire before it runs down its target. On Africa's short-grass plains – typical cheetah habitat – cover is in short supply. So a cheetah will spend 10–20 minutes stalking, head lowered in a semi-crouch, creeping forward, or running and then freezing before edging cautiously forward again, finally launching into its explosive run.

Unlike hyenas and African wild dogs, which select their prey during the hunt, a cheetah begins by stalking one animal and rarely switches target during the chase. Its energetic lifestyle means it needs to prey on fairly large animals. But having the body of a sprinter, it lacks the strength to fight its prey to the ground as a lion or leopard would. Instead, it uses its long dewclaw to trip up its prey. When the prey loses its balance and rolls over, the cheetah clamps down on the windpipe with its jaws and kills by strangulation. Apart from size, the most important factor for a cheetah when selecting prey is the animal's vigilance. Cheetahs abandon 75 per cent of their hunts after being

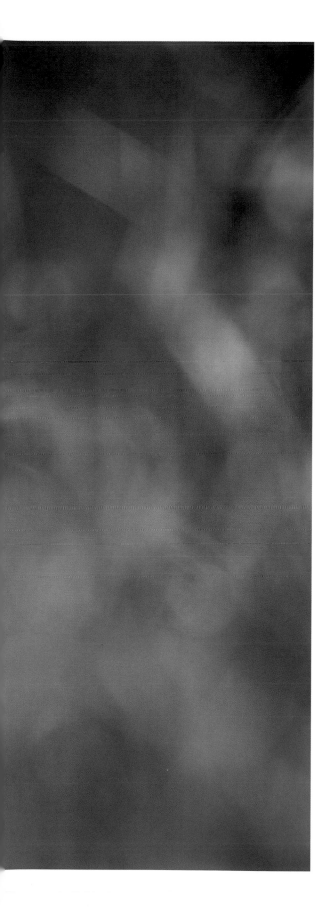

detected stalking. Inexperienced gazelle fawns are therefore a favourite prey. Less vigilant individuals on the edges of groups of Thomson's gazelles – usually males – also tend to be caught.

When it launches its sprint, a cheetah's breathing rate rockets from 60 to 150 breaths per minute and its heart production increases more than 50-fold. But speed alone is not enough because antelopes are adept at quick zigzag turns. Only from the air can you really appreciate how manoeuvrable a cheetah is. Its hind claws stay open through the chase, providing grip, but its front claws are partially retracted, keeping them sharp for ripping up the prey.

Of all of Africa's predators, cheetahs are second only to wild dogs in their hunting success. On average, 50 per cent of their hunts result in a kill. But the downside of being designed for speed is that a cheetah can't defend itself or its kills from lions and hyenas. In the Serengeti, more than 13 per cent of cheetah kills are stolen.

THE MASTER OF STEALTH

You can spend months in the African bush and never even glimpse a leopard, and few have succeeded in filming one hunting in daylight. A leopard's spotted coat is perfect camouflage for the dappled vegetation in which it likes to skulk. Master of the stalk, it uses cover more than any other predator and relies on getting extremely close – within 4 metres (13 feet) – before grabbing an animal with a short burst of speed. This is an adaptable strategy, which has been successful throughout sub-Saharan Africa, India and southern Asia, in alpine mountains, arid desert, open grassland and tropical rainforest.

Leopards also have a broad taste – in Africa alone, 92 species have been recorded as prey. They are big enough and strong enough to handle animals as large as an eland but will go for smaller prey, particularly in more difficult times. The perfect leopard prey – impala, bushbuck and duiker, for example – weighs about 20kg (44 pounds), but where the pickings are thin, leopards will take prey as diverse as porcupines, peafowl and monkeys.

◄ **The watcher.** A leopard watches impalas grazing in Kenya's Massai Mara. Its strategy is to stalk to within striking distance (within 4 metres) of mainly medium-sized prey such as bushbuck, impala and duiker, which can be dragged into cover or up a tree away from lions and hyenas.

A leopard is a master at keeping close to the ground, moving slowly and silently. Vegetation as low as 20cm (8 inches) is enough for concealment. But the highest density of this skulking predator is found in the dense tropical forests of central Africa. Here the vegetation provides ample cover, and the leopard's strategy changes to hunting by ambush, often for duikers (smallish antelope) or monkeys attracted by the bounty dropped from fruiting trees. Prey as large as red river hogs and as intelligent as chimpanzees have been caught in this way. In fact, leopards are a serious threat to chimpanzees, which have been seen ganging up and cornering a leopard, threatening it with broken branches.

A leopard's most effective cover is darkness, and in most parts of its range, it is almost entirely nocturnal. Under the revealing light of an infrared

The master stalker. Keeping close to the ground, leopards can hide behind vegetation just 20cm (8 inches) high. But even so, daytime hunts are rarely successful, and throughout most of their range, leopards usually hunt at night.

camera, it becomes obvious that it is master of the stalking hunt as you see it inching closer to its prey. Though every rustle of leaves or the crack of a twig seems so much louder than in daylight, the gazelle remains oblivious and keeps chewing until the leopard bursts forward. There's a last-second alarm call from the antelope, a cloud of dust and then silence.

A leopard's hunting success is, though, far lower than a cheetah's. In the Serengeti, only 5 per cent of stalks are likely to end up with a kill. In the denser vegetation of South Africa's Kruger National Park, the success rate rises to 16 per cent, and in Namibia, where leopards are almost entirely nocturnal, it is 38 per cent. Even if they do make a kill, leopards lose up to 10 per cent to lions and spotted hyenas. But unlike cheetahs, leopards have the strength to pull their kills up into trees, out of the reach of their competitors.

GANGING UP

Many predator species choose to hunt in groups, usually because it helps them to catch larger prey. Though three or four cheetah brothers or sisters may hunt together for at least part of their lives, among the big cats, only lions live and hunt together their whole lives. The size of a lion pride varies with the quality of its territory, ranging from 4 to 40 – usually females with their cubs and one or two adult males, who protect the pride but do little of the hunting. Only when a pride goes for larger prey such as a buffalo or elephant will a male join the hunt, mainly because a male weighs almost twice as much as a lioness

▼ **Success in numbers.** A large pride of females sharing a wildebeest. With so many mouths to feed, the territory needs to be rich in prey. In fact pride size is less important for hunting than for defending territory and fighting off hyenas and rival lions that might kill their cubs.

and is slower. Indeed, lions are not designed for speed – the top sprinting speed of a lioness is less than 64kph (40mph), and they lack stamina. On the flat, a lioness can run for only two or three minutes, which means that lions' favourite prey – zebra, wildebeest and buffalo – can easily outrun them. So, to have any chance of success, lions need to get within 30 metres (100 feet) of their prey. This means using tall vegetation as cover or hunting at night.

In open landscape such as Etosha National Park, they also work together to set a trap, some of the pride driving prey into the jaws of others, lying in wait. The chances of success then increase directly in proportion to the numbers of lionesses involved. But in this harsh habitat, even the most successful prides only kill 15 per cent of the time.

In the Serengeti, where the density of prey is much greater, the success rate increases to 23 per cent. But here the advantage of group hunting is less clear. Lionesses hunting together and sharing the kills do not appear to get more meat than a solitary one would hunting on its own. So why do most of the Serengeti's lionesses still hunt in prides? There seem to be two key reasons. First, large prides are better able to defend the best territories, which in the Serengeti tend to be the wooded areas near the rivers where the game feels more secure. These prides are more successful, and the females rear more cubs. The second advantage is protection from rival lions and other predators such as hyenas that might steal cubs. Overall, lionesses in prides enjoy greater reproductive success than their solitary sisters.

THE RELENTLESS PACK

Africa's most impressive social predator must surely be the African wild dog. The hunting success rate of a pack is rarely less than one in three and can reach as high as 85 per cent. It is hard to think of any other predator that kills with such efficiency. African wild dogs always hunt together, and at times, a pack can be as large as 40–50. Though they weigh in at about 25kg (55 pounds) in East Africa and 30kg (66 pounds) in the south, to have any chance of bringing down prey as large as zebra and wildebeest, the dogs have to work as a pack.

An African wild dog cooperative hunt is one of the most dramatic you will ever see, though even today, the true nature of the cooperation is not fully understood. Like cheetahs, they are visual hunters, and for a time it was

thought they only hunted in the day, usually around dawn or dusk. Today we know that both species will hunt at night if there is enough moonlight. All wild dog hunts start with a noisy greeting session, with lots of whining and twittering calls, the dogs excitedly wagging their tails, licking and rubbing up against each other. This reinforces bonds and re-establishes the hierarchy before the hunt begins. Then they head off following the lead dog. They need to warm up before the chase. They start by walking and then shift into a trot. Finally, they switch to a steady, loping run. Now they are ready for the hunt.

Once the lead dog spots suitable prey, it will stop to let the others catch up. As they arrive, they spread out, focusing intently on the prey. Then, with heads held low, the dogs start to stalk. Only when their prey takes flight does everything shift gear. Reaching up to 64kph (40mph), the frantic chase is on.

In the woodland of southern Africa, these chases may only go on for about 600–800 metres (655–875 yards), and the lead dog will stop the hunt if there seems to be little chance of success. But on the open plains of East Africa, they may last for up to 20 minutes and cover well over a kilometre. Opinions differ on how cooperative this chase really is, but look down on a hunt from the air, and it certainly seems as if the dogs are working together.

▲ **Marathon runners.** On the plains of Zambia, African wild dogs keep a fast pace in pursuit of wildebeest. Their strategy is to tire out their prey.

▶ **Selecting the target.** The large pack closes in on the wildebeest. On this hunt the dogs successfully isolate and eventually pull down their target animal.

In their woodland habitat, you will often see a single dog diverge from the pack and seem to try to cut off the escape route of the prey. And watching a more sustained hunt out in the open, it is clear that the role of the lead is handed from one dog to another as the chase progresses. Some believe this relay gives the whole pack more stamina, ensuring the lead dog is always fresh. Others wonder why the second dog that takes over should be any less exhausted than the leader it replaces. But when they finally come to kill, the real value of hunting as a pack is obvious.

A single wild dog couldn't bring down an adult wildebeest on its own, but a pack can. Having separated a wildebeest from the herd, the dogs take it in turns to bite at its back legs and try to disable it. Eventually, two or three dogs will be able to grab its tail and slow it down. The final stages can seem gruesome as they rip into their prey, but it is also when the true nature

and value of the pack becomes clear. All the wild dogs enjoy the spoils. The hunters use a distinctive 'hoo' call that carries long distances through the woodland to bring any lost dogs into the kill. Unlike other social hunters, only when the younger dogs have had their fill do the older ones eat. The pups of the year eat first, while the adults spread out around the carcass, looking for hyenas or lions that might come to steal their prize.

Spotted hyenas are three times the size of an African wild dog, and it takes four adult wild dogs to face off a hyena from a kill. But even the whole pack can't see off a single lion, and lions are the biggest normal cause of death in wild dogs. The second most important cause is broken legs. The hunt is so fast and frantic that many wild dogs damage their legs in the chase. By living and hunting together, they can at least ensure the risk of these fatal injuries can be mitigated by the pack.

▼ **Dog power.** A hyena is driven off by a pack of African wild dogs after attempting to take their kill. Though the hyena is much larger than an individual wild dog, it's no match for a pack, which may even be able to keep several hyenas at bay. But should a lion appear, the dogs won't risk doing battle with the much larger and more powerful rival.

HUGE TRAPS AND TINY TRAPPERS

When it comes to the smaller solitary hunters, subterfuge and trickery are often employed, and in the tropical rainforests, camouflage and mimicry are common weapons in the arms race between predators and prey. One of the most impressive traps is that of Darwin's bark spider. This Madagascan weaver, the size of a fingernail, is the only known spider to build its web over a river or a lake. A river provides an insect flight path through the rainforest and therefore can be a good place to set an invisible trap. The challenge is how to get it there.

Darwin's bark spider uses silk that is the toughest known natural material – twice as strong as any other spider's silk and ten times stronger than the high-tensile man-made Kevlar. She also builds one of the largest known orb webs (only females build webs), with anchor lines stretching up to 25 metres (82 feet) across the water. The spider starts her work in a well-positioned bush on one bank of the river, and from her spinnerets at the end of her body shoots out tens of light silk threads, designed to catch the wind and carry across the river. Once one of these strands becomes entangled on vegetation on the other

▶ **Bridging the river.** The giant web of a Darwin's bark spider hangs from bridging lines across a Madagascan river. The orb web, which can be more than 2.5 metres (8.2 feet) wide, is positioned to catch insects using the river as a flight path and is renewed daily. Like the bridging lines, it is exceptionally strong, able to catch large, fast-flying prey.

▼ **Firing the silk.** A 2cm Darwin's bark spider shoots threads of silk from the spinnerets at the end of her body. The silk catches in the wind and carries across the river. One reason for its high-tensile strength is its elasticity and the fact that tens of strands form the bridging 'rope'.

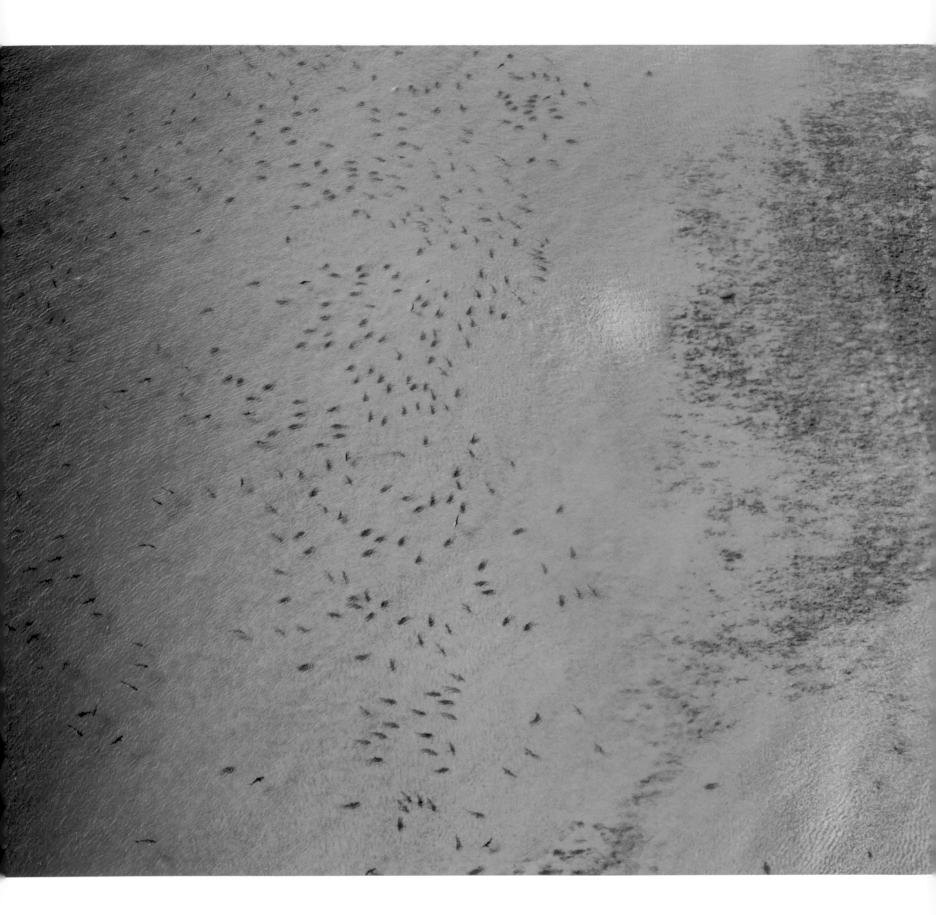

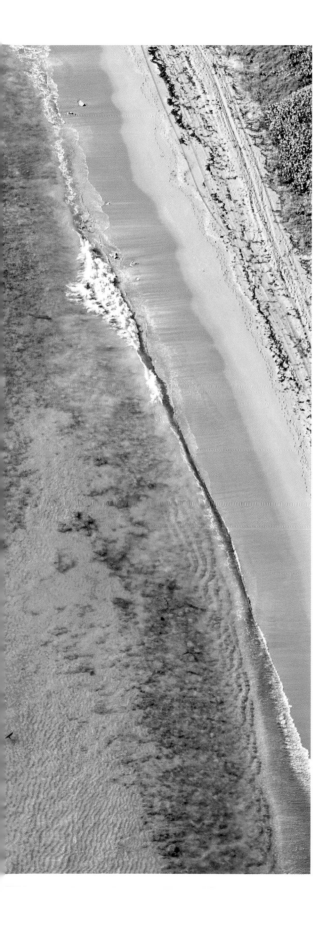

bank, she pulls it tight and makes her way out on it. She may spend up to a couple of hours strengthening this vital bridging strand and checking that it is safely secured on both banks. She then descends to create a third attachment point. With this in place, her last task is to construct the orb web at the central T-junction of the three bridging strands. Up to an hour or so later, the trap is set.

This construction is a major investment of time and energy. The silk is particularly elastic, which seems to protect it from snapping in the wind and tropical rainfalls, but even so the spider will replace the bridging strands regularly and renew the central catching-web every day. Her industry is appreciated by others. Male Darwin's bark spiders steal food from it, and there is even a fly that specializes in feeding on the smaller insects that become entangled in the web.

HUNTING ON THE MOVE

Twice a year, billions of animals migrate vast distances across the globe following the sun on its six-month cycle, and many predators have no option but to follow them. These migrations create some of the greatest spectacles in the natural world, though not all are well known.

Every autumn, large schools of black-tip and spinner sharks migrate to the coast of Florida, escaping the northern winter, then moving back again in the spring. Swimmers are unaware of them, but from the air, thousands of sharks can be seen just a few hundred metres or so offshore. These sharks are themselves followed by their own predators – larger hammerhead and bull sharks.

Two hundred different birds of prey migrate thousands of miles each year. Most rely on thermals to help carry them on their journeys and so prefer to migrate overland. Where the land narrows, birds of prey may be concentrated in spectacular numbers. In Veracruz, Mexico, 5 million pass overhead each year, crossing between North and South America – a spectacle known as the river of raptors.

◄ **Sharks on the move.** Thousands of black-tip and spinner sharks migrate north along the coast of Florida, having wintered farther south. Sea temperature may trigger the migration, and the movements of prey such as mullet may be the reason for it.

▶ **(next page) Falcon stopoff.** After a mammoth journey south from their East Asian breeding areas, tens of thousands of Amur falcons stop off in northeastern India to feed up on insects before a more than 3000km (1865-mile) journey to southern Africa.

▲ **Falcon rest stop.** Amur falcons (males and females) roost on a telegraph wire in Nagaland, northeastern India – one of the mid-migration stopoffs on the way to wintering areas in southern Africa.

◄ **Refuelling on termites.** Amur falcons feed beside Nagaland's Doyang reservoir. They are just some of the tens of thousands that time their stopover to coincide with the emergence of termites. Local people now regard this spectacle as a valuable tourist attraction rather than an opportunity for hunting. In the past, 120,000 or more falcons were killed annually at stopovers in the province.

The longest annual journey by a European raptor is a 14,485km (9000-mile) round trip undertaken by steppe buzzards from northern Europe to South Africa. But the longest and most arduous of all raptor migration routes is that of the Amur falcon. This kestrel-sized falcon breeds in the eastern Palearctic (including areas of Russia, Mongolia and south to central China). It specializes in catching insects on the wing, and when the insects start to die off in autumn, the falcons have no option but to head south to where insects are plentiful. This takes them on a 22,530km (14,000-mile) round trip to southern Africa.

Some cross over the Himalayas in central Nepal, but most avoid such high altitudes by flying along the eastern edge of the Tibetan Plateau. When they reach northeastern India, they stop for several months to refuel and build up fat. Tens of thousands hunt together for termites in dense flocks reminiscent of roosting starlings. Finally they are ready for the most demanding part of their journey – a non-stop flight of more than 3000km (1865 miles) across open water to Africa. This is the longest open-ocean migration by any raptor and takes the Amur falcons two or three days. Their prize after this endurance test is a rich supply of insect prey and the long days of a southern summer.

◄ **Ringed seal haul.** A polar bear drags a ringed seal from the sea, having caught it in the water as it dived off the ice and before it could swim down out of reach. When the ice breaks up in summer, this staple prey animal becomes far more difficult to hunt, and polar bears enter a season of hunger.

► **The sea bear.** Using its huge forefeet as paddles and its hindfeet as rudders, and insulated by a layer of fat and a dense undercoat of fur, a polar bear swims across open water in search of ice where seals may be lying out. In summers of extreme sea-ice retreat, when a bear is forced to cross large areas of ice-free sea, it may have to swim for days non-stop.

THE PREDATOR FOR ALL SEASONS

The polar bear is one predator that doesn't migrate with the seasons. It is such a specialist hunter that it is entirely restricted to the frozen sea ice of the Arctic. Nevertheless, the Inuit of northwest Greenland call it *pisugtooq*, which means the wanderer. A polar bear's home range can be vast – the average size is as large as the US state of Georgia, with the largest recorded being almost the size of Texas – and it is constantly on the move, in search of seals. But it faces one massive challenge, unique to the polar regions: every six months, the sun's arrival melts much of the polar bear's ice world. To survive, it has become by far the most adaptable predator on the planet.

No other predator has evolved so many different hunting techniques to deal with the changing seasons. As spring turns to summer, polar bears need to take up a different hunting technique almost monthly. Early in the season, their favourite prey are the pups of ringed seals, which are born in lairs under the sea ice in March and April. But when that ice starts to break up and the

ringed seal pups are weaned, bears turn to bearded seals. By autumn, when all the ice has gone and with it the seals, the bears may even resort to attacking walruses to steal their pups.

The life-or-death balance for a predator is between energy expended hunting and energy acquired. Physiologically, a polar bear is extraordinary. It has a digestion that assimilates 84 per cent of the protein and 97 per cent of the fat that it consumes, and when walking out on the ice in winter in search of food, its metabolic rate can drop to as low as that of a black bear 'hibernating' in its den. This is just one of the adaptations that makes it possible for the polar bear to survive in extreme conditions.

▲ **Hard times.** A male polar bear climbs along a cliff ledge in search of guillemot eggs, risking a fall into the sea far below. Melting sea ice has made seal-hunting impossible and has forced the desperate bear onto this island in the Siberian Arctic.

THE ULTIMATE PREDATOR

Perhaps the ultimate predator, in terms of intelligence and adaptability – and success worldwide – is the killer whale. It has speed, strength and stamina. It catches its prey with a level of cooperation and intelligence unmatched in any other hunt. No ocean animal, however large, is beyond its reach.

Killer whales can live for more than 50 years, and they maintain stable matriarchal family groups for decades. Cultural differences between groups are passed down from generation to generation. The more we get to know about this highly social marine mammal, the more we realize how intelligent it is. Only humans are more complex and adaptable.

There are distinct populations of killer whales, which are referred to as ecotypes, though genetic indicators suggest that they may need to be classified as different species. They look different, have different acoustic behaviour and hunt for different prey in different ways. Their ranges hardly overlap, and they are not thought to interbreed.

The first type to be recognized as distinct lives in the Pacific off the North American Northwest. These killer whales are referred to as residents, staying around the many shallow bays that line this long coastline. They are fish-eaters, with a preference for the chinook salmon that return in huge numbers each year. Their pods are 80–90 strong, all close kin, and they stay together throughout their 50-year lives. Underwater communication is a vital

▲ **Fish-chasing.** Fifty or so resident killer whales, including young calves, chase a huge school of herrings in a Norwegian fjord. These killer whales specialize in hunting fish in the cold, nutrient-rich Scandinavian seas.

▲ **Herring-herding.** Frantic herrings are herded up against the surface and into a ball. The killer whales are using echolocation to locate and drive the fish and to communicate with each other as they round them up. Tail slaps disable the fish before they are eaten.

part of their hunting coordination, and researchers have been able to identify each resident pod by its distinct dialect. In fact, the variety of killer whale dialects seems to be without precedent in any other non-human mammal.

The other type of killer whale hunting along the west coast of North America is referred to as transient because they are always on the move. The same individuals have been seen as far south as southern California and as far north as the Bering Sea. One group satellite-tagged off Alaska travelled 1400km (870 miles) in just 8 days, right up to the edge of the Arctic sea ice. Transients specialize in hunting mammals rather than fish, and they have an extraordinary ability to turn up in the right place at just the right time to encounter their prey.

WHALES HUNTING WHALES

Each year grey whale mothers and their young calves leave the nursery lagoons off the coast of Mexico to start the long migration north to the rich feeding grounds of the Bering Sea. They can't go fast because of the calves, but by April most reach Monterey Bay on the California coast. The shortest route north is to cut across the bay, but this is where pods of transient killer whales are waiting.

These transients are far less vocal than residents, probably to avoid their presence being heard by their marine mammal prey, which have similar underwater communication systems. It would be too risky for the killer whales to attack a grey whale mother, which is a lot larger and has extremely thick skin and thick blubber. Instead they go for her calf. Only by working together do the killer whales stand a chance of separating it from its mother, and only the large adult females are involved in the hunt. They have to be careful to avoid being injured by the thrashing of the mother's powerful tail, and once they succeed in separating the calf, they have to try to drown it by leaping onto its back. The frantic pursuit can take two to six hours.

Another favourite ambush spot for the transients is along the coast of Alaska and the Unimak Pass, which the migrating grey whales reach by midsummer. When a pod attacks, the mothers head for shallow water, often nearly grounding themselves to avoid the killer whales. Despite this, 5–15 per cent of the calves are taken as the migrating grey whales travel through the Unimak Pass on the way to the Bering Sea.

Near this ambush point is another rich hunting ground for the mammal-hunting killer whales. Thousands of fur seals breed on the Pribilof Islands, and every year the transients arrive in late May to coincide with the start of the fur seal breeding season. They mainly target the young male fur seals, fat and ready for sex but forced out to the edge of the colony. Then in the autumn, the killer whales have an opportunity to prey on the pups leaving the rookeries.

◄ **Transient attack.** A killer whale (right) rams into a grey whale calf, trying to separate it from its mother and drown it. This killer whale is one of a transient group that preys on marine mammals off the Pacific coast of North America. The family pods ambush grey whale calves in spring as they migrate with their mothers up the coast, in particular – as here – in Monterey Bay National Marine Sanctuary, California.

1

Attack of the baby-killers

Killer whales that specialize in hunting large mammals such as seals and whales will work like a wolf pack, using stealth, cooperation and strategy. One such population ambushes humpback whales migrating from their Antarctic feeding grounds to their tropical breeding grounds.

The killer whales arrive off the coast of Western Australia in autumn to prey on early-born calves. The humpback mothers hug the coast trying to avoid being detected by the killers. But once the killer whales detect a mother and calf, they will pursue them. Successful attacks may last just a few minutes or hours. The mother is too large for the killer whales to drown, but her baby is small, has little stamina and can't hold its breath for long, and so it has to rely on her for defence or on the assistance of other whales. Most of the time, the killer whales are successful, and approximately two thirds of the attacked calves are killed and eaten.

▲ **1. Race for life.** The mother humpback has lifted her calf onto her back, temporarily out of reach of the killers. Ahead, one of six attacking killer whales tries to block the female. The outriders are two male humpbacks who are defending her and the calf, hitting out with flukes and flippers, blowing bubbles and trumpeting – making a smokescreen to block the killers' vision.

▶ **2. Riding on mother.** Exhausted, the baby lies on its mother's back, while she tries to outrun the pack of killers.

3. Lashing out. Desperate to fend off the killers, the mother lashes out with her tail and fins, formidable weapons that, helped by an encrustation of barnacles, can kill.

4. Closing in. The killers have managed to separate the calf from its mother, and they can now force it down from the surface and drown it. But they will have to feed fast, as the battle has attracted sharks, which will crowd in to scavenge on the kill.

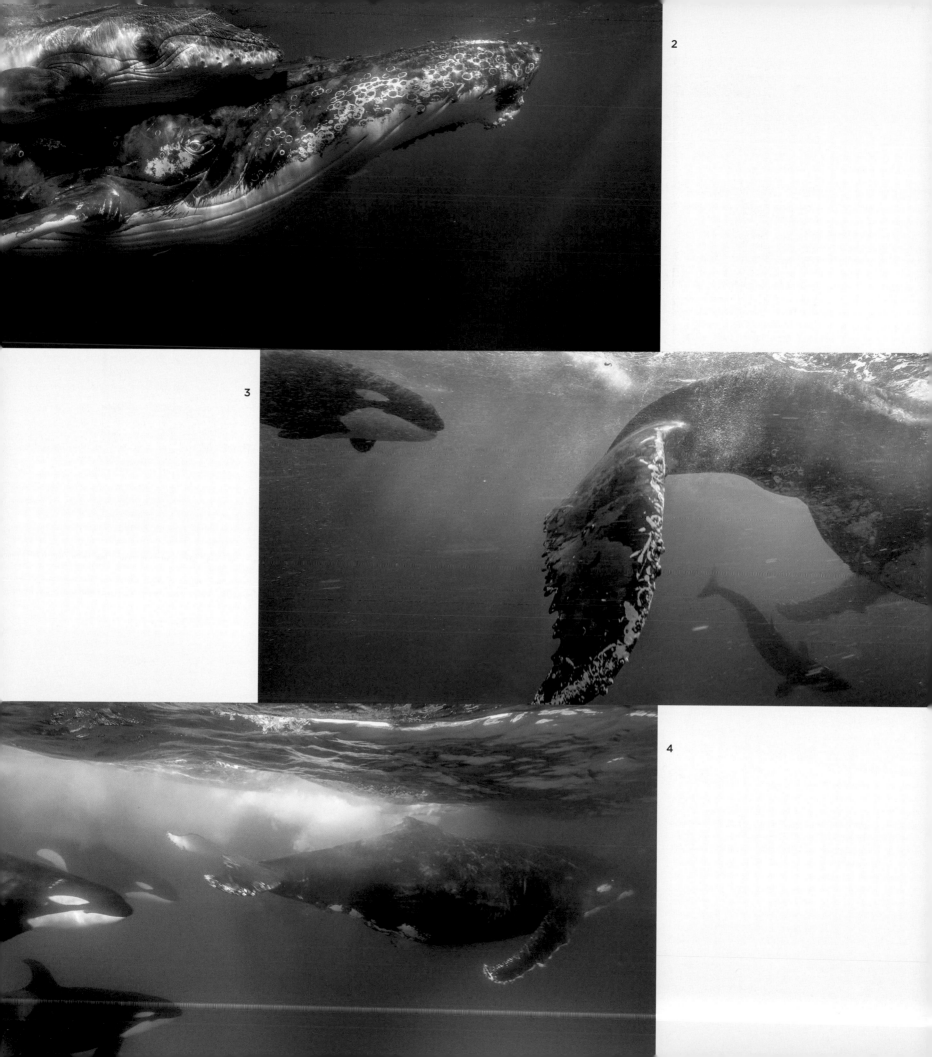

2

3

4

WHALES HUNTING SHARKS

The third ecotype of killer whale found off the coast of North America is called offshore because they hunt far from the coast along the continental shelf. They are smaller and faster moving and form larger pods of about a hundred strong. They are also elusive, and for a long while their diet was a mystery. They tend to have extremely worn teeth, which suggested they might be biting shark skin (the embedded denticles in shark skin are so abrasive that it was once used as sandpaper). Then researchers off Alaska found shark liver at the surface near offshore hunting areas. It seems that these killer whales are

deep divers, submerging for up to five minutes on the hunt for the deep-water Pacific sleeper sharks, whose livers are rich in energy – a prize well worth a little wear and tear of teeth.

THE CLEVEREST KILLERS

Researchers have recently started to uncover a fascinating new range of hunting techniques used by killer whales in the Antarctic. One ecotype, known as type A, keeps mostly to the Southern Ocean's open waters and specializes in killing the minke whales that are common here. Type Cs are found only in east Antarctica, and at just 6 metres (20 feet), they are the smallest of all the killer whales. Each spring, as the sea ice breaks up, they swim up cracks into shallow water, where they hunt for Antarctic toothfish.

The largest of all the killer whales, type B, penetrate deep into the pack ice all around the continent. Within this ecotype, there are two distinct hunting techniques and groups of whales. The smaller of the two have been seen hunting along the Antarctic Peninsula, where they go for penguins and probably fish as well. The larger of the type Bs specialize in washing seals off ice floes. This is a rich food source, because Antarctic seals are thought to be among the most numerous of all mammals.

Once they have found a seal, the killer whales approach to take a closer look. If it is a crabeater seal, they usually move off and find another target. It seems the crabeater's sharp teeth and feisty nature makes it a challenge even for these top predators. If it is a more docile Weddell seal, the pod lines up and swims towards the ice floe as a team. Just before they reach the ice, they dive forcefully down, creating a powerful wave that crashes over the floe. The seal can only cling on terrified as the killers repeatedly wash the floe with waves and break up the refuge into ever-smaller pieces. Eventually the killer whales get close enough to spyhop and size up their seal prey.

The seal backs itself into a crevice in the ice and snaps at the killer whales with its sharp teeth. Once again the killers' teamwork comes into play. Taking it in turns, the whales use violent sweeps of their tails to try to wash the seal out into open water. If that doesn't work, they might resort to blowing blasts of bubbles to drive out their prey. The seal rarely escapes, and no one who has witnessed the killer whales at work can be left in any doubt that these are the planet's ultimate wild mammal predators.

▼ **The last move.** Two killer whales check the position of a terrified Antarctic Weddell seal before making their final move: a second formation-charge to create a wave that will tip the ice and wash the seal into the water. The ice has already been broken up by the first wave they created, but the seal managed to climb back onto the remnant.

CHAPTER TWO
HIDE AND SEEK

RAINFORESTS SUPPORT THE GREATEST
DIVERSITY OF LIFE – on one tree alone there
can be more than a thousand species of insect.
But walk into a rainforest for the first time, and
you realize that the animals that live here are
experts in the art of not being seen. There are
notable exceptions, such as birds of paradise,
but in the main, tropical forests are full of highly
camouflaged animals, both predators and
prey. And in a world of trunks, branches, lianas
and leaves, that's not too difficult. Even where
the understorey is relatively open, visibility is
limited, not helped by the fact that only about
2 per cent of sunlight filters down to the forest
floor. So for predators, just finding prey can be
difficult. The vegetation in seasonal forests is
less dense, but visibility can still be a problem,
and so it's a hide-and-seek situation here, too.

▶ **Rainforest invisibility.** A tangle of buttress roots, branches and understorey
vegetation hides thousands of small animals, both predators and prey.

◀ **(previous page) Dry-forest hide-and-seek.** A tiger stalks a chital deer in
India's Bandhavgarh National Park, blending in with its environment.

THE ALL-SEASON HUNTER

Playing hide-and-seek with an American pine marten would be a very long game indeed. Able to move effortlessly between the forest floor and the treetops in the northern temperate and boreal forests, martens are a 'here one second and gone the next' kind of animal – their fast metabolic rate resulting in a frenetic lifestyle. The continuous need for calories means these solitary hunters spend much of their waking hours scouring their forest home for a meal. In the middle of summer, when food is most plentiful, they may put in regular 16-hour shifts.

American martens do best in undisturbed forests, where the ground is more likely to be littered with dead wood and branches and where their favourite prey is to be found. These lithe carnivores – mustelids, as are stoats and weasels – have a broad diet, including hares, squirrels and birds, but voles are what they like most, to be found among the litter on the forest floor.

There's a long list of forest predators capable of catching voles, but the American marten is arguably the most adept of them all – and it's all down to physique. The marten's long, thin, flexible body and short legs are great for climbing trees and chasing squirrels in the branches, but they're also the perfect features for hunting small prey through tight gaps, crevices and tunnels. When a marten is on the trail of a vole, it seldom escapes. But having the ideal vole-hunter's body shape comes at a price. A long, narrow body means a small,

◀ **Fast-food hunter.** Speedy and ever-curious, an American pine marten forages in leaf-litter on the forest floor in New England, USA. Its high metabolism means that its search for food is constant, up trees and down holes, with voles top of the menu.

◄ **Snow-hole shelter.** A marten emerges from its bivouac. Its lifestyle means that it can't put on enough fat to hibernate in winter, but it does use snow holes for shelter from extreme cold. It also hunts for voles and mice under the blanket of snow, though without body fat to fall back on, it has a daily struggle to balance its energy budget: food intake versus activity. So larger prey such as snowshoe hares become worth hunting in winter.

► **Vertical lookout.** Equally agile in the forest canopy, a martin searches cavities and bark for roosting birds, insects and squirrels. Trees also provide refuge from predators such as foxes, and in winter, evergreen conifers provide overhead cover and the coarse, woody forest-floor debris where rodents can be found.

narrow stomach. So the marten can't gorge itself when the hunting is good, and therefore has limited fat reserves to fall back on when the long northern winter sets in. A long, narrow body also has a high surface area to volume ratio, which means a marten loses heat much faster than if it were wide and squat with a low surface-to-volume ratio. The only way to generate heat, as with all warm-blooded animals, is to burn calories. That means converting voles into energy. This isn't a problem in summer, when prey is usually abundant, but in winter, temperatures in the boreal forests can drop as low as −20°C (−4°F). And with no fat reserves, a marten can't hibernate as some other mammals can and has to continue to find food somehow.

To survive winter, an American marten does have some useful adaptations. One is a dense, luxuriant coat, which partly compensates for its large surface area. And when temperatures plummet dangerously, it's able to seek shelter in snow holes, much like a polar explorer, dropping into a shallow torpor to save energy. Deep snow also has little impact on a marten's movements, since its large furry feet help to spread its weight, allowing it to almost bounce across the surface. But the snow does add another layer of protection for the marten's prey, making hunting even more of a challenge.

During winter, martens have to become predators of the subnivean world – the world under the snow – in particular the spaces that form around

low-hanging branches, logs and other woody litter, places where voles remain active. Wherever martens find vegetation poking out from the snow, they stop, as these could provide access points to subnivean spaces. Then they use their strong senses of smell and hearing to detect any voles – and on average, one in ten holes will test positive.

Since voles are normally much scarcer in winter, martens will also seek out carrion – perhaps animals that have succumbed to the cold. But larger predators such as coyotes and foxes also seek out these carcasses, and they would see a marten as competition if not a meal. Its only escape would be upwards. Indeed, scientists now believe that the marten's tree-climbing skills are more an adaptation for eluding predators than chasing arboreal prey.

THE SWOOP-AND-STOOP HUNTER

You don't often think of birds of prey being henpecked, but it's an appropriate description for the male sparrowhawk. The female can be nearly double the weight of the male. This size difference – one of the largest of any bird – is so great that they look like different species, and it means that it's the female who rules the roost when they pair to rear young. During this period, the male tries to keep her happy with food offerings, but without getting too close.

After the eggs hatch, he provides food for the growing family for two weeks, which means bringing up to ten kills a day. He is not built for stamina, but in short bursts he can fly at speeds of up to 50kph (31mph). Surprise is key. He sneaks up on his prey – mainly small birds such as tits and finches – concealing himself in the woodland foliage, and when he breaks cover it's a short, swift chase. Then his small size comes into its own, since it gives him much greater manoeuvrability among the branches and foliage than his huge mate. By folding his short, rounded wings at just the right moment he can shoot through the narrowest gaps while chasing his jinking prey through the understorey.

It's not easy singling out a target, and, in fact, only one in ten attempts end in a capture. His songbird prey may stay in groups for safety, and at the first sign of danger they will try to take refuge in foliage too dense for the sparrowhawk to follow. His best success is when ambushing young, inexperienced birds. It's why sparrowhawks time their breeding to coincide with the fledging of their prey – so that when their own chicks, and the mother, are at their most demanding, there is a bounty of potential targets.

▲ **Canopy ambush.** With a target in view, a male sparrowhawk dives through the canopy, fast and silent. Short glides alternate with quick, deep wingbeats, its closed tail acting as a rudder. It has a 10 per cent chance of catching its prey.

▶ **(next page) Practice strike.** A young male attempts a strike. At virtually the same size as the jay, it has no chance of being successful but is improving its strike skills. Had the sparrowhawk been an adult female – much larger than a male – the jay would have already flown.

Each kill needs to be handed over to his partner carefully. He takes the meal to a feeding post and waits for the female. As she flies in, he makes a sharp exit, leaving the carcass for her. These brief handovers are one of the few moments when you can appreciate the impressive size difference between the sexes.

By the time the job of fatherhood is over, the male sparrowhawk is faced with another problem – the onset of winter. Bare trees offer fewer opportunities for concealment, and the tits, finches and other small songbirds now move around in larger flocks for added safety, making it much harder for him to sneak up on his prey. The female, being bigger, can go after larger prey such as jays, magpies and woodpigeons. For her, the lack of cover may actually be an advantage. For him, life is tough, and he'll be lucky to see out four years.

STEALTH, CAMOUFLAGE AND POWER

There are few more impressive animals than a wild tiger. If you're ever lucky enough to see one in its natural habitat, such as the sal forests of central India, the image is likely to be burnt into your memory. This largest of all cats is the planet's most formidable forest predator. A male Bengal tiger can take down an adult gaur, a huge wild cow up to six times heavier than a tiger. But its main prey – chital, sambar deer, wild pigs, langur monkeys – actually make

▲ **Under cover.** Stalking through elephant grass in India's Kaziranga National Park, a male tiger reveals how its stripes can be the perfect camouflage for an ambush predator.

little attempt to hide in the forest. Instead, these animals rely for safety on their acute senses and security in numbers. Indeed, park guides often locate tigers by listening to the sound of alarm calls – the *ack-cack-ack* of a langur or the piercing *aawwww* of a chital, or spotted deer. The calls not only warn other members of the group about the threat, they also serve notice on the tiger that its game is up.

For a successful hunt, the tiger needs to keep completely out of sight. This is when its beautiful striped coat comes into its own. In the dappled light of

the forest floor, or even the long-grass edge of a meadow, the combination of gold and black breaks up its outline, and it can melt into the vegetation. This is perfect for a sit-and-wait ambush, but a tiger can't always rely on its prey coming to it and so must be able to play both sides of the hide-and-seek game at the same time. It's a challenging task, so knowing the best places to look can help enormously.

A female tiger has a home range of about 20 square kilometres (4940 acres), small enough for her to gain a good knowledge of her territory – paths regularly used by her prey, waterholes that they like to visit, which places they like to forage in and when. But knowing the hot spots is only half the job. She also has to learn where to hide before launching an attack. It's a level of familiarity that could make the difference between survival and starvation. A male tiger holds a territory more than three times the size of a female's – just too large to get to know well. So adding that to the stress of fighting to defend his territory, it's perhaps not surprising that his life expectancy is several years shorter.

When hunting, a tiger relies on sight and hearing. Like all cats, it has excellent night vision, and forward-facing eyes enable it to accurately assess distance and depth – useful when manoeuvring through the 3D world of a forest. But hearing is a tiger's most acute sense. Rotating like radar dishes, its ears pick up the faintest sounds of prey and allow it to 'see' around bushes and tree trunks.

Tigers can reach speeds of about 65kph (40mph) over short distances, but like many forest predators, they have to get close to their prey before they break cover. The last few metres of the stalk may take 20 minutes or more, and a paw may hover in the air for what seems like an age. But the longer the stalk, the more chance there is of the prey detecting the tiger or just wandering out of ambush range. Indeed, only one in twenty tiger hunts ends in success.

But when the prey is in range, powerful back legs can launch the tiger several metres in a single leap, and death comes quickly as the largest canines of any big cat quickly sever the prey's neck. A meal of a deer will provide food for several days, but to survive in these monsoon forests, the tiger has

◄ **Group alert.** A tiger stays motionless, knowing that she is not within striking distance and that the chital deer have spotted her. Foraging in a group multiplies the deer's chances of spotting a tiger, their main predator. The high-step warning posture of several of them is a signal to the tiger that they are aware of it and are fit and able to flee.

to repeat this success at least 50 times a year – and more often than that if a female has cubs to feed.

Rearing cubs will test a tigress's hunting skills to the limit. She could be supporting as many as four cubs, which means catching a large deer every four days, and the cubs will remain dependent on her until they're about 18 months old. She'll start teaching them the basics of hunting even before they've been weaned at six months, but they won't be able to kill until their canines are fully developed at about 14 months. And while she's showing them the ropes, they're likely to spoil a lot of her hunts, either by making a noise or appearing at the wrong time.

For cubs, learning how to catch and kill prey is trial and error, and it's an art they will still be learning even after their mother has called time on their schooling. It's easy to understand why fewer than half survive to adulthood.

KILLING IN THE CANOPY

The harpy eagle is the largest and most powerful bird of prey in the Americas and the top aerial forest hunter. It was named after the harpy of Greek mythology – a winged monster with sharp claws and the face of a woman. The female harpy eagle can be double the weight of a male, with a great hooked bill and talons the size of a man's hands – the biggest of any living eagle. Scientists wanting to put telemetry tags on chicks in their nests are well advised to wear riot-control gear (helmets, stab vests, thigh and shoulder padding) in case of attack by the female – a wise precaution when dangling on a rope high in the canopy.

But being huge requires a considerable amount of food, and finding and catching the right prey in the forest is a challenge, evidenced by the time it takes a pair of harpy eagles to raise the next generation. While a sparrowhawk chick is independent after about 50 days, a harpy chick can take up to 24 months. This huge investment in time is unique among birds and significantly longer than in the majority of mammals.

The single chick (harpies never have more than one) fledges after six months, but it will need another year, or more, to hone its hunting skills – and it can only do that if it's regularly supported by its parents. The adults may even avoid hunting prey directly around their nest to give the chick something to practise on. But when the 12 months or so of practice are up, the parents

▶ **The true harpy.** A female harpy eagle rests close to the nest, displaying her enormous feet and talons, designed to grab and dispatch a sloth or monkey. Her breast feathers are stained from nearly a year of catering for her chick, which will remain at the nest for many months yet.

aggressively drive the chick out of their home range. It's the harpy's way of saying, 'this territory isn't big enough for the three of us' – and there's an even greater sense of urgency if the parents are about to raise another chick.

With its relatively small wingspan – an adaptation that increases its manoeuvrability among the branches of the canopy – eyesight that's up to eight times more powerful than that of the average human, and excellent hearing – enhanced by its disc-like face which helps to focus the sounds – the harpy eagle is superbly adapted to its rainforest habitat. Its main problem is the same as for most forest predators: finding prey among the tangle of vegetation and then sneaking up on it without being seen. Once again, it's hide-and-seek, a job made more difficult by the kind of prey favoured by harpy eagles.

The diet of the harpy eagle varies from one region to another and among individuals and may include peccaries, agoutis and armadillos, as well as birds and reptiles. But its favourite prey is generally a tree-living mammal, such as a sloth or monkey, which is why it is sometimes referred to as a monkey eagle.

Sloths, with a slow-motion lifestyle and exceedingly cryptic coloration – a result of their algae-coated fur – are hard to spot among the mass of leaves and branches. Monkeys are even more challenging: fast moving, intelligent, highly visual and always in groups, which means plenty of eyes on the lookout for danger. They can also be dangerous. Their strong arms and tight grip – not to mention teeth – could easily damage a harpy's delicate wings. Hunting success, therefore, requires time and patience. Surprisingly, given its size, there is little first-hand knowledge of a harpy eagle's hunting behaviour. But then, given its habitat, the size of its territory and that it is monitored from the ground, it's perhaps not so surprising.

To see a successful hunt you need to have the luck of the forest gods. There are rare eyewitness accounts of harpy eagles snatching howler monkeys out of the treetops and carrying them up and over the forest – no mean feat, as a full-grown howler can weigh as much as a female harpy. Such sightings invariably occur around forest gaps, such as on the edge of rivers or lakes, where it's possible to view the canopy. But over the years, observations by scientists, naturalists and film-makers have begun to build up a picture of their strategies.

◀ **Mother protector.** Crouched over the remains of a sloth, a mother harpy eagle dismembers it to feed to her chick, which she shelters from the daily downpour. The nest is high up in the canopy, and there are no predators that will threaten the giant chick.

Harpy eagles almost never hunt by soaring over the canopy and don't stoop on prey. Their approach is mostly one of stealth and ambush, hiding in the canopy, waiting for something to come along. Once a harpy has a target, it goes into stalk mode until close enough to launch a surprise attack, just as a jaguar would on the forest floor. It attacks from behind – something scientists working around harpy nests know well. Those that have been attacked always say the same thing – they didn't know it was there until it struck their backs. Once caught, prey will be carried to the forest floor where it will be stabbed with the eagle's 13cm (5-inch) rear talons – claws larger than those of a brown bear.

Some experts believe that harpy eagles are capable of building up a picture of all the primates in their territory – assessing which monkeys in a group are dangerous and which are potential targets. And there's no doubt they spend a lot of time monitoring potential prey in their home range. Occasionally an eagle may even hunt monkeys by drawing attention to itself, calling from a hidden position and then listening to the result. If there's little or no alarm reaction from the troop, then they're obviously not eagle-aware and so an attack is worth risking.

FEROCIOUS LITTLE JUMPERS

On the micro level, an even bigger game of hide-and-seek is taking place. Forests offer infinite places to hide, and nothing sums this up better than one study in Amazonian Ecuador that discovered literally thousands of species of beetles on just one species of tree. Among the invertebrate predators, the most common are probably spiders. In every rainforest, there are likely to be hundreds if not thousands of species. Many of them set traps – invisible webs – and wait for passing insects to be caught in the sticky threads. The jumping spiders, though, practise a stalk-and-ambush strategy, which can involve tackling prey many times bigger than themselves.

Among the most astonishing of the jumping spiders are the *Portia* species. The white-moustached *Portia* from the forests of Asia adapts its hunting techniques according to the prey and is so versatile that it's been called an

▶ **Jumping a spitter.** A *Portia* spider with her fangs in a spitting spider. She has crept up on it from behind, partly to avoid the spider spitting venomous sticky silk at her, and then jumped on it, delivering the fatal bite. Her huge eyes indicate her acute vision.

eight-legged cat. Its main prey are web-building spiders, which it catches by stalking them across the threads of their webs. The challenge is to do this without the owner of the web becoming aware. If it is too heavy-footed, the target will beat a hasty retreat or possibly turn nasty. To counter this, *Portia* has a number of tricks. If there are breezes shaking the web, it will time its movements along the strands of silk to coincide with the vibrations. If the web owner does get suspicious, then *Portia*'s camouflage may make the poor-sighted web spider think a piece of leaf detritus has got caught.

Sometimes a *Portia*'s strategy is to attract the would-be victim by mimicking the struggle of a trapped insect or the courtship signals of a male – repeating any pattern of movement that succeeds in getting the prey to move closer. This can take a while, but she appears to have unlimited patience. During one observation, a *Portia* spider spent three days vibrating a web before finally getting a response.

If things aren't going to plan and the web owner begins to act aggressively, then *Portia* will retreat and rethink its strategy. It might take a detour and attempt to come at the target from a different direction to retain the element of surprise. It's an extraordinary level of sophistication for an invertebrate hunter.

But perhaps *Portia*'s most challenging hunt is for spitting spiders. These kill their prey by spraying a liquid mixture of venom and silk, which congeals on contact. So *Portia* has to attack from behind – unless the spitting spider has its mouth full of eggs, when *Portia* will risk a frontal attack. It's another example of the spider's hunting intelligence.

CARNIVOROUS LITTLE PRIMATES

The cover of darkness gives prey yet another layer of protection. It's the reason why so many small, vulnerable forest animals are nocturnal. Predators that hunt at night therefore need extra-keen senses. In the case of Southeast Asian tarsiers, they have the biggest eyes of any mammal of their size. Each eye is as large and heavy as a tarsier's brain, giving these small, social primates exceptional night vision. They are also the only completely carnivorous primates. Though they eat mainly insects, such as katydids – jungle crickets – they will also catch lizards, small snakes and even birds.

A tarsier's hunting strategy is best described as a leap and a bite. Its powerful hindlegs are like coiled springs, able to launch the pocket-sized predator up to 5 metres (16 feet) from a stationary position. In addition to oversized eyes and

◀ **Night-jumper.** A Philippine tarsier munches on a cricket, caught in a matter of seconds and consumed on the spot. Tarsiers are the only totally carnivorous primates. Their night-hunting adaptations include highly sensitive hearing, swivel ears, the largest eyes proportionately of any mammal – able to pick up the slightest light at night – and the ability to hear ultrasound.

▶ **Emerging for the hunt.** Spectral tarsiers emerge from their daytime resting place in the rainforest of Sulawesi, Indonesia, to start hunting. Using their extremely long legs as springs, they leap both to catch prey and to move between trees. Their fingers and toes are also extremely long, each with a pad to help with grabbing and gripping.

springy legs, other adaptations include being able to turn its head nearly 180 degrees in either direction – effectively giving it a 360-degree view, handy for checking the surroundings without moving a limb. It also has large, sensitive ears that pick up the slightest noise – such as the movement of an insect.

It has recently been revealed that a tarsier can even hear and vocalize in ultrasound – rare among land mammals. It's a discovery that has given the Philippine tarsier the accolade of having the world's highest-frequency primate call. It's like having its own private channel for communication while keeping under the audio radar of both prey and predators.

But the tarsiers' small size – none are bigger than about 16cm (6.5 inches), not including the tail – means that they, too, are vulnerable to other nocturnal predators, such as civets, owls and large tree-climbing snakes. This could explain why some tarsiers live in small groups. Not only does this mean more eyes looking out for danger, but as a team, tarsiers can gang up and mob a potential threat such as a python.

HUNTING-PARTY PLANNING

The popular perception of the chimpanzee, our closest relative, is that it's a highly social, intelligent fruit-eater. But though it has a plant-based diet, it's also a predator – of other primates. It acquires this extra protein by forming hunting groups, behaviour almost unheard of among rainforest mammals.

In the Ivory Coast's Tai Forest, the main prey is red colobus monkeys. A single chimp would struggle to keep up with a troop of colobus travelling through the canopy, and if it did manage to get close, the much lighter monkeys would retreat to branches too slender to support an ape. So, to have any chance of success, the chimps need to set a trap.

The average number in a chimpanzee hunting party is four or five, though it can be as many as ten or as few as two. With two, the chance of success

Hunting-party handouts. After a successful red colobus monkey hunt, the chimpanzee hunters share out the meat among themselves, while others in the community watch and wait expectantly. Individual hunters will then give away bits to favoured individuals. Such sharing can help create and cement friendships. Females may also choose to mate with males who are generous with their meat.

▼ Sharing with mates. A favoured female about to be presented with a monkey leg by one of the hunters.

drops dramatically, but even with double that number the success rate may be just one in three. Success also depends on experience. Male chimps – and it's always the males – start learning the art of hunting at around six years old but don't become proficient until they're thirty. And monkey-hunting is only worth the effort in the rainy season, when the difficulty of gripping slippery branches makes the colobus more cautious about jumping from tree to tree and less willing to hang onto frail vegetation. It's also when the females have their young, which makes them, and the group, much less mobile.

What cues the chimpanzees to begin a hunt isn't clear, but when they do, silence falls. With an unspoken understanding, several chimpanzees start to climb into the trees. One will act as the driver, while the rest go ahead to act as blockers or ambushers. When all are in place, the driver begins to move towards the monkeys. With so many eyes looking out for danger, it doesn't take long for the colobus to spot him and flee. They don't always head in the predicted direction, and so the blockers and ambushers must alter their positions accordingly.

If all goes to plan, the troop of colobus will come within striking distance of one of the ambushers, who will grab a smaller, weaker monkey. Only one chimp makes the catch, but all members of the hunt share in the spoils. A successful hunt causes the forest to echo with the screams of excited chimps. The hunters decide who gets what – mostly determined by status. This has led some scientists to conclude that hunting and meat-sharing is less of a foraging necessity than of a social act, designed to build relationships and allegiances.

THE MARCHING ENTITY THAT EATS ALL IN ITS PATH

There is one kind of rainforest predator that takes cooperative hunting to the extreme: the army ants of Central and South America. What they lack in size they make up for in overwhelming numbers – colonies can number more than a million. In one sense a colony is a giant superorganism, which in a single day can harvest a staggering 30,000 prey items, making it one of the most successful predators on the planet.

A raiding party of *Eciton burchellii* (the most studied of the 200 species of army ants) is an impressive sight. A swarm can be several metres wide and 200 metres (655 feet) long. If you step into one – perhaps because you've been looking for wildlife up in the canopy – the first indication is likely to be a stabbing pain in your leg as a soldier ant cuts into your flesh with its scythe-like 'jaws'. Indeed, these mandibles are so formidable that some Amerindian tribes have used them as stitches to close wounds. And it's not just the soldiers that inflict pain: the workers have stings. As the ants sweep across the forest floor, the leaf-litter fizzes with fleeing creatures. If ever there was a moment to get a proper appreciation of the incredible diversity of the rainforest, then this it. All kinds of small creatures are on the move – scorpions, spiders, crickets, beetles – each trying to keep ahead of the ravenous horde. Not many do.

An army ant raid is conducted with military precision, though without a leader, with the thousands of workers coordinating their behaviour and responding to each other through touch and chemical (pheromone) communication. The integration is all the more impressive because the ants are blind. While the column drives forward, the ants maintain three lanes of traffic, one inner and two outer. The inner lane comprises the food-carrying ants, which take the shortest route back to their base camp, following pheromone trails laid by earlier ants. The outer lanes comprise ants moving to the swarm's front, also following pheromone trails. Having separate lanes keeps outgoing and incoming ants from bumping into each other.

Prey is detected mainly by vibration. The slightest movement attracts the ants like iron filings to a magnet. Small prey is killed by a sting, larger prey is overpowered and dismembered. Even huge tarantulas – formidable predators in their own right – stand little chance against the coordinated attack. But bigger rewards come from the nest of another social insect. Overpowering a colony of wasps or ants is no mean feat. Clashes between army ants and Azteca ants can last 40 minutes or more – and the action is as dramatic as any Hollywood battle

▲ **Tentacle of a superorganism.** Workers in a raiding column march off into the forest guarded by a 'major' army ant, recognizable by its ivory-coloured head and massive, sickle-shaped piercing 'jaws' (mandibles). As soon as the first prey is caught, an inner stream of workers will form, flanked by outgoing ones, each carrying or helping to carry parcels of prey back to the base-camp nest.

scene. The advantage can swing from one side to the other, with members of both species being killed. But usually the relentless pressure and power of the army ants win through, and they claim the spoils – the Aztecas' larvae. Every grub in the nest will be taken and carried back to the ants' bivouac.

The one defence strategy against the ants that has some chance of working is to keep completely still, taking advantage of the ants' blindness. To do this requires nerves of steel… or being a stick insect. Even when ants nibble their legs, stick insects will remain stationary. Another trick, adopted by some spiders and caterpillars threatened by an approaching swam of ants is to dangle out of range on silken threads. The least successful strategy is to flee.

The efficiency of numbers

Army ants hunt by swarm, attacking, subduing and dismembering a wide range of prey species. Half are likely to be other ants and half large arthropods (spiders and insects – crickets, for example). If the prey stays still, it has a chance of escaping death. But once it moves, the ants will sting it.

Accompanying the medium-sized workers are the much larger submajors – the porters. A worker can carry back a small food item, such as an ant, but a submajor deals with larger stuff. She will also dismember larger prey, as her huge jaws can cut through tough cuticle, though group effort may pull apart big animals at the joints. A submajor also has longer legs so she can carry large items under her body. But if something is too big for her to carry on her own, a group of workers will gather to help. Their numbers are precisely organized to match the weight of the item. Food is carried at a set marching speed (presumably to keep the ant highway running smoothly), with the ants' leg movements coordinated for the most efficient load transport.

▶ **1–4 Death of a cricket.**
Eciton burchellii worker ants swarm over a cricket that has failed to leap beyond their path. Arching their abdomens, they insert their stings. They are joined by submajor ants **(3)**, which use their larger jaws to start cutting into the cuticle, while workers tug at the limbs to pull them from their sockets. Once the cricket is dismembered, groups gather, each with a submajor, to haul away the pieces.

FOLLOWERS AND FORAGERS

For those that do escape the ants, it's often into the mouths of other predators. Indeed, more than 500 species are known to benefit from the ants, including tamarin monkeys, which snatch insects as they flee, a predatory beetle that mimics the smell of the ants and runs alongside them, and parasitizing flies that lay eggs in the heads of fleeing insects. The most noticeable followers are the antbirds. Some, such as the ocellated antbird, are so dependent on the ants for a meal that they would starve without them. To get enough food, they monitor the whereabouts of several colonies of army ants at the same time, as army ants don't forage every day and colonies frequently emigrate to new nest locations.

Army ants, such as *Eciton burchellii*, have two distinct phases, nomadic and stationary. The stationary phase lasts 20 days, when the colony becomes

▲ **Daily follower.** A spotted antbird follows its daily routine of travelling alongside army ants, picking off insects fleeing from the terrifying swarm. This species isn't totally dependent on the ants to flush out food, but 20–30 other birds are frequently to be found with the ants or are specialist ant-followers.

▲ **The food-transport team.** A porter worker ant carries part of a dismembered insect back to camp. She has grasping mandibles and extra-long legs, and carries her awkward load slung beneath her. She also has a team of two small helpers from the basic-worker caste. On either side can be seen the legs of outgoing workers on their way to the front of the army.

an ant-rearing factory, with the queen pumping out as many as 100,000 eggs. During this period the ants may not hunt every day. When they do it's always in a fresh section of forest, as they shift their foraging direction out from the bivouac nest site. This ensures that they are not hunting in areas that have only recently been harvested.

In the nomadic phase, which lasts 15 days, the ants raid every day, sweeping a new section of forest and advancing roughly 100 metres (330 feet). This behaviour also allows the raided sections to recover their prey densities. It's the only way the ants can keep up with the massive demands of the colony.

Indeed, measured by the weight of the prey they consume, these ants have a bigger impact on the forest than jaguars and, in superorganism terms, they could be said to be the biggest predator in the forest.

CHAPTER THREE
NOWHERE TO HIDE

THE PHRASE 'NOWHERE TO HIDE' CONJURES UP THE NIGHTMARE of fleeing from a pursuing killer with no way of escape – the stock scene of horror movies. For prey animals living on the plains, grasslands and deserts of the world, the nightmare is real. Plains herbivores such as antelopes don't have anywhere to hide, which is why so many grazing animals opt for safety in numbers. Such gatherings include the million-strong trains of wildebeest migrating between the grazing areas of Tanzania's Serengeti and Kenya's Massai Mara and the aggregations of ground-nesting birds such as snow geese. Having nowhere to hide applies to the predators, too. Even if the target animal hasn't spotted the hunter, one of the others in the group probably will have. To survive in such open environments requires particular strategies.

▶ **Plain numbers.** Wildebeest in search of fresh grass. In a landscape where there's nowhere to hide, safety in numbers is the strategy for most grazers.

◀ **(previous page) The ten-second sprint.** A cheetah in mid-air full extension, on a trajectory that anticipates the swerve of the Thomson's gazelle.

SPEED KILLS

No animal looks better designed for a life on the open plains than the cheetah. Its legendary speed and acceleration – a cheetah can go from zero to 30 metres (98 feet) in a single second, faster than any racing car – is ideally suited to this environment. It can outpace prey in a sprint, though success may depend on agility – something a cheetah also excels at. But such specialization has a cost. Having a lithe, slender body means that a cheetah can't defend a kill from lions, hyenas or even vultures.

Lack of cover makes it difficult both to stalk prey to within sprinting distance and to hide a kill from the competition. Even the presence of a tree is of little help as, unlike leopards, cheetahs can barely climb, let alone stash prey out of reach in branches. A cheetah will drag its kill into the cover of bushes or long grass. But hyenas have acute smell and hearing – they are thought to be able to hear the crunching of bones from several kilometres away – and vultures, with their almost supernatural senses of smell and sight, are quick to detect death on the plains. Also, both lions and hyenas are known to use circling vultures to locate carcasses. When either of these two formidable predators appears, there's little a lone cheetah can do except abandon its kill.

In some locations it's thought that up to 30 per cent of cheetah kills are lost to stronger competitors, though the average might be closer to 15–20 per cent. This is probably one reason why cheetahs hunt more often in the hotter

◀ **Racing pose.** Perfectly built for speed on the flat, a cheetah intently watches gazelles grazing nearby. Vigilance is key to knowing when to start a sprint. Its profile shows its long, flexible back, long, muscular legs and front-foot dewclaw used for tripping prey.

part of the day, when larger rivals are more likely to be crashed out in the shade. Cheetahs eat quickly, too. It can take less than two hours for a cheetah to consume an adult Thomson's gazelle. The coalitions formed between cheetah brothers to hold territories could also make it easier to defend kills.

Having nowhere to hide cubs also affects cheetah survival. Across parts of its range, less than 5 per cent of cubs make it to adulthood. The causes include being killed by other predators, especially lions.

How to move undetected when there's so little cover and exactly where to launch the chase – usually within 30 metres (98 feet) of its target – is something a young cheetah struggles to learn. Even if it becomes able to hunt for itself, a cheetah's life expectancy is not high. The average for a male is less than three years and for a female just over six.

▲ **The struggle.** A cheetah tries to hold down a large male Grant's gazelle that she's caught but is lifted off the ground as it makes a powerful leap. The gazelle got away, though with a broken leg, and it was a week before the cheetah had a successful hunt.

▶ **The standoff.** A family of cheetahs tries to fend off a hyena. But where there is one hyena, there will soon be more, and it isn't worth risking injury trying to defend their food from the clan. It's possible, though, for a coalition of adult brothers to defend a kill if there aren't too many of the bullies.

◀ **The escape route.** Having taken refuge from hyenas, a young caracal considers whether to climb farther up a tree. Though mainly a hunter of ground-feeding birds, hares and rodents, it's an opportunist hunter and, if necessary, will climb for a meal, using its strong, curved claws.

▶ **The catch.** A caracal holds down a red-billed francolin it has just caught with a bound and a leap as the bird tried to fly up. The caracal used both its front paws to snag it.

LEAPING FOR A LIVING

If you were to say or do something that causes trouble, you could be said to have 'put the cat among the pigeons'. The origin of this phrase is actually a reference to the caracal – a wild cat from Africa and Asia – and its amazing agility when hunting. In Iran, trained caracals used to be put in arenas with flocks of pigeons, and wagers were made as to how many the caracal would knock down. The record was apparently 12 at once.

Aside from its wonderfully tufted ears and golden coat, a caracal's most distinctive feature is its powerful hindlegs, noticeably taller than the frontlegs. This enables it to leap up to 3 metres (10 feet) when trying to hook a bird in flight. It's also the fastest cat for its size, able to run down prey such as hares. And unusually for a small cat, it can kill prey such as small antelopes two or three times bigger than itself. So a caracal can be an opportunistic hunter and target whatever prey happens to be the most abundant at the time.

Caracals favour more open, arid country than African wild cats, though they also need some cover so they can stalk as close as possible before rushing out and pouncing. Their hearing is acute: their ears work like parabolic aerials, pinpointing the exact location of prey.

SMELL IT, DIG IT UP, EAT IT

Being a predator doesn't necessarily exempt you from the menu of other, bigger predators. But a honey badger – listed by the *Guinness Book of Records* as 'the world's most fearless creature' – will successfully stand its ground against much larger hunters. For defence, it has very strong claws and a thick hide, particularly around the neck, where an attack would normally be directed. If something grabs it from behind, its loose skin allows it to turn and face down its assailant. It also has backup – an anal pouch that can create a suffocating smell. Even lions will think twice before trying to take on this smelly, muscled mustelid. Some scientists have even theorized that the silver manes sported by cheetah cubs may have evolved to mimic the honey badger's coloration. Indeed, young cubs sometimes seem to move much like a foraging honey badger.

▶ **The digging machine.** Displaying its massive claws and muscular neck and shoulders, a honey badger pauses at the entrance to its den.

▼ **Copy cats?** Young cheetahs display their cub coats, which some say resemble the coloration of a honey badger. The theory is that this might give them a little protection from predators that have already had experience of the badger's ferocity and skunk-like defensive smell.

The honey badger, or ratel, has a largely 'see it and eat it' diet. In the Kalahari, in southern Africa, it is known to prey on more than 60 different species – from beetle larvae, scorpions and snakes to rodents, lizards and birds – and will climb trees after bees' nests and happily raid beehives – hence its name. It also has a high metabolic rate and therefore a particularly large appetite.

A honey badger will eat, on average, about a kilo of food a day – though one 11kg (24-pound) male, was recorded eating more than 6kg (13lb) of meat in a day, including four adult mole snakes, two adders and seven mice. He wasn't making up for a lean week either: the previous day he'd consumed more than 2.5kg (5.5lb) of meat, and on the day after, he ate another 3.3kg (7.2lb).

A honey badger's greatest hunting tool is its sense of smell, which more than makes up for its poor eyesight. Much of its prey lives underground, and its hunting strategy can be summed up as: smell prey, dig it up, eat it. In one day a honey badger might dig 50 holes and cover a distance of more than 40km (25miles). It also has tricks. A honey badger hunting for rodents has been seen digging alternately at two or three holes while trying to block off potential exits with its hindfeet.

Venomous snakes such as puff adders are more of a challenge, but a badger invariably gets its prey, even if the snake lands a successful strike – it seems that honey badgers are largely immune to snake toxins.

Though mostly nocturnal, a honey badger is also often active in the cooler parts of the day, and when food is limited, such as during the Kalahari's cold, dry season, it will spend even more time foraging in daylight, making little effort to hide. It has no real reason to keep a low profile, being without serious predators, and there's one animal that has exploited this fact. In southern Africa, pale chanting goshawks have learnt to follow foraging honey badgers. From the vantage point of a rock or bush, the goshawk will swoop on any prey that escapes the badger's clutches. It is a remarkably successful strategy, accounting for up to 60 per cent of escaping prey. Good for the goshawk but of no benefit to the badger.

◀ **Snake-killer.** Having dug down into a hole, a honey badger struggles to pull out a huge puff adder, which it has by the tail. It has no fear of snakes, usually killing them with a bite to the head, and is believed to be immune to the venom. In search of grubs, it will explore any hole, crevice or mound, where it often encounters resting snakes.

THE DEADLY LIGHTHOUSE

Millions of years before the first humans walked the planet, another animal came up with the defence strategy of a castle or fort. Go to almost any tropical plain or grassland across the world, and the chances are that you'll see mounds of baked earth rising from the ground. Some are surprisingly large – the record is more than 12 metres (40 feet) tall – a size all the more impressive given that the architects are little insects – termites.

There are about 2600 species of termite. All are plant-eaters and all live in colonies that can number several million strong. They are also one of the most protein-rich foods in the world, recorded in the diet of more than 130 species, including humans. So it's not surprising that these soft-bodied insects have learnt to protect themselves by living within a rock-hard home – one that can withstand fire, rain and most predators.

The only animals that can break into termite mounds are specialists such as aardvarks and giant anteaters. Both species have claws like grappling irons, which they use to rip open the walls, and long sticky tongues that can reach into the mound's network of tunnels. A giant anteater's tongue is more than 50cm (20 inches) long, and it can flick it in and out of its toothless jaw 160 times a minute, enabling it to lap up tens of thousands of termites a day – though not in one sitting. As soon as the termite mounds are breached, soldier termites attack the intruder with a ferocity that's hard to ignore. After ten minutes, most predators move on.

But there is another termite predator that has a very different strategy. The Brazilian cerrado grasslands are home to the greatest concentration of termite mounds in the world, castles of clay often more than 2 metres (6.5 feet) tall. Examination of the rough surface of an old mound reveals tiny holes. These are the tunnels of predatory click beetle larvae – also known as headlight beetles – and there may be as many as 400 living on a single mound.

Each larva creates a U-shaped tunnel in the outer wall of the mound, where it will stay until it's time to pupate. At one end of the tunnel, the gallery

▲ **The cerrado light show.** All over the Brazilian cerrado, just after the rainy season, the termite mounds glow with the lights of click beetle larvae. (The light in the sky is that of an adult beetle, caught in the long photographic exposure.) They turn on the sinister light show only on nights when the termites' nuptial flight happens and the air is full of flying insects.

is enlarged to create a side chamber for storing termite food. The problem it faces is how to catch the termites without entering the fortress or leaving the safety of its burrow. The answer is both sinister and spectacular.

The chink in the termites' armour is the need to emerge from the fortress to mate and start new colonies. After the annual rains, when the ground is soft enough to burrow into, reproductive winged termites – known as alates – are released from the mound in their thousands. The alates are the future queens and their suitors, and the aim of each queen is to start a new colony. Very few succeed. Predation is so intense that it's estimated that less than 0.5 per cent of queens survive. Birds snatch the termites out of the air, and frogs and lizards snap up those that land on the ground. The predatory click beetle larvae take a more patient approach, but then they may have already been waiting ten months for this opportunity.

Most alates emerge around dusk. This is when the click beetle larvae poke out of their burrows and start shining light into the darkness, produced from a gland in the thorax. The mounds start to glow with hundreds of little green lights – bioluminescence on a grand scale and a truly otherworldly spectacle.

The flying alates are lured to the mounds by the green glowing lights. As soon as one lands in range, the larva strikes – gripping the termite with vice-like jaws and dragging it back into the tunnel to store in the larder. Then it it goes straight back for more. Alates only appear for a few weeks a year, and so for the click beetle larvae, this emergence might be the last chance of food until the next rainy season. Even then it takes at least two seasons for each larva to catch enough food to mature into adulthood.

BEATING THE NUMBERS GAME

When there's nowhere to hide, one of the best defensive strategies is safety in numbers, and there are few examples more impressive than the great flocks of snow geese that gather each spring and autumn at Squaw Creek, Missouri, in the USA.

Squaw Creek's lakes are a major stopover for the geese as they head north to the Arctic and their breeding grounds, and when they migrate south again in October. At their peak in early March there may be well over a million birds, and the mass take-off as the snow geese leave their roosting area each morning must rank as one of the greatest natural history spectacles.

▶ **(next page) A blizzard of snow geese.** Thousands of geese lift up from their night-time roost in Missouri, USA, to feed in the surrounding area. It would be a bold bird of prey that would risk damage diving into such a flock. So for these hunters, another tactic is required.

An appropriate collective noun for such a flight would be a blizzard of snow geese. To pick a goose out of this vast aerial crowd would need a strong, skilled and determined predator, and there are none capable. But on the sidelines there are hunters waiting for a different kind of opportunity.

Each year, up to 300 bald eagles (occasionally more) gather at Squaw Creek in March and October – tough times for the mainly fish-eating eagles. But they have a problem. A healthy snow goose weighs about 3.5kg (7.7 pounds) and is too large and strong for a bald eagle to attack, and the eagle is more skilled at using its powerful beak and talons to catch large fish.

But the eagles have a snow goose hunting technique that is both simple and successful. The more geese there are, the more likely the strategy is to work, and the colder the weather, the better the hunting. The short length of the Arctic summer makes the migration of the snow geese very finely balanced. If they reach the Arctic too early, the ground will be covered with snow and unsuitable for laying their eggs, but too late and they might run out of time to raise their young. Weather conditions at Squaw Creek, a midway point in their 4000km (2485-mile) migration, can be a good indication of

▲ **Fly-past scare tactics.** A bald eagle makes a low pass over a huge flock of snow geese looking for weak or injured birds. There is also the chance that some might panic, collide and break wings or legs and become easy prey.

spring's progress. When the geese arrive at Squaw Creek in late February, the lakes may or may not be iced over – though the situation can change in a single day. For the hungry bald eagles, it's freezing conditions that give them an advantage.

The eagles' technique is to fly low over the flock resting on the water, which scares the geese into the air. They are looking for weak or injured individuals. There is also the possibility that, as the panicking birds take off, some collide. Collisions can lead to broken wings or legs that makes a goose easier for an eagle to tackle. When the lakes are free of ice, the strategy is less successful as the geese can spread out and injured birds can duck beneath the surface to escape the swooping eagles.

An injured goose on the ice will usually put up a vigorous defence, but unable to escape, the bird is gradually worn down by the relentless attacks of one or more eagles. Once dead, a carcass will attract as many as a dozen eagles, and there will be a lot of jostling and posturing in an attempt to get a slice of the action. Bald eagles are skilled scavengers, and few will go away empty-taloned.

Nothing to fear – yet. On the Alaskan tundra, an incubating snowy owl pays no attention to a greater white-fronted goose on its way to its own nest close by. Highly aggressive in defence of their own nest, the owls will keep out foxes and gulls and create a safety zone for the goose pair. The protection payment comes when the goslings emerge.

Chick-fodder time. A female owl flies to her nest with a lemming for the newly hatched owlets. Once goslings become available, they will also be on the menu.

WHEN PREDATORS PROTECT PREY

In May on the Alaskan and Canadian tundra, pairs of greater white-fronted geese have to make a vital decision: where to position their nests. On the open tundra, there are no trees or bushes – nowhere to hide – and so they have no option but to nest on the ground. But the choice of where to nest can still make the difference between breeding success and failure.

A raised hummock or slope may reduce any flooding risk and on relatively flat tundra can provide a view of the surrounding area and therefore predators – birds of prey, Arctic foxes and gulls. Male geese will aggressively defend their nests, and a pair working together can often keep predators at bay. But there is another kind of defence that can keep the eggs and goslings safe, at least for a time.

Studies have shown that some pairs of geese will choose to nest within view of snowy owl nests – sitting as close as 10 metres (33 feet) away. Newly hatched goslings make ideal food for snowy owl chicks, and so why risk nesting close to the owls? It could be that the advantages more than make up for the disadvantages. A snowy owl pair – particularly the male – will aggressively defend their eggs from predation by foxes and gulls and will dive-bomb them repeatedly if any come within 500 metres (1640 feet) of the nest. Any adjacent geese incubating their clutches benefit from the banishment of the egg thieves. It's a kind of protection racket, and though the owls may not be complicit in the arrangement they will eventually claim their dues.

Snowy owl eggs begin hatching around the middle of June – usually a week or two ahead of the goose clutches. If it's a good season, an owl may raise as many as 11 chicks, which require hunting for round-the-clock. The favourite prey is lemmings, which provide most of the chicks' nutrition. When lemmings are plentiful, the owl parents stack the nest with these tundra rodents. By the time an owl chick has reached independence it's likely to have eaten more than 150 lemmings. But goslings are on the menu, too.

Unlike the owls, which are asynchronous layers (eggs are laid two days or more apart and therefore hatch at intervals), the goose clutch hatches within 24 hours. It's now that the owls begin to take notice of their noisy neighbours. While the goslings are in the nest, with the parents sheltering them, they are largely safe from the adult owls, which don't tend to recognize something as food unless it moves. But a day after the chicks have hatched, the parents have to herd them down to water to feed.

The goslings are able to walk, swim and feed as soon as they hatch – unlike the owlets, which are born blind and helpless – but they are totally dependent on their parents for protection. On the journey to water, the goose pair will try to keep the goslings together, but inevitably some chicks lag behind. A gosling on its own, even for a moment, is an easy target for an owl. It's only when they reach water that the goose family is safe from aerial attack.

But the window for owl predation on goslings is relatively short. Every goose nest will hatch within a couple of weeks, and they will gang up together for added protection, with the female geese herding the chicks, and the males defending. With more eyes keeping a lookout for danger there is less chance in this open environment of a surprise owl attack.

▶ **A larder of lemmings.** Snowy owl chicks rest, surrounded by a store of dead rodents. Despite such a glut of lemmings, the parent may still find their neighbours' goslings irresistible once they hatch and set off to find water.

THE PRANCING WOLF AND THE BOBBING MOLE RAT

The Ethiopian wolf has a different kind of hunting challenge. Its favourite prey – giant mole rats – only appear above ground briefly each day, and so a wolf has to remain on alert. Living on the roof of Africa, in Ethiopia's Bale and Simien Mountains, the wolf's home is a combination of flat plains and open valleys. So while on the one hand its view of the terrain is mostly unobstructed, there's little in the way of cover – at least, little that's higher than a long-legged wolf.

Ethiopian wolves are social animals and live in packs, but this is principally for pup-rearing. With prey animals not big enough to share, each wolf must hunt alone. At around 30cm long (12 inches), a giant mole rat is not large, and though it is a rodent and does appear rather rat-like, albeit a weird-looking one, it is neither a mole nor a rat. But like a mole, it spends most of its life underground and much of its time tunnelling. It pops up out of a new hole only to pull vegetation down into its tunnel to consume in its own time, and is never above ground for more than 20 minutes at a time and no more than about an hour a day. Even then, the mole rat always keeps its legs in its burrow, ready for a speedy retreat. So catching one is a challenge.

An Ethiopian wolf has keen eyesight and can spot a mole rat from a distance, but with nowhere to hide, getting close is another matter. Different wolves use different techniques. Some just run and hope for the best – though unsurprisingly this is not a successful strategy. The mole rats are very sensitive

◀ **The prancing wolf.** After patiently waiting for a mole rat to pop up from its hole, an Ethiopian wolf pounces. If the mole rat bobs down in time, the wolf will blow down the hole and, guided by the noise, will start frantically digging into the tunnel in advance of its retreat.

▶ **The bobbing mole rat.** A mole rat pops up to check if it's safe to start cropping the surrounding vegetation and pulling it down into its tunnel. Being mainly subterranean, it has small eyes and ears – all the better for the wolf – but it spends as little time as possible above ground, feeding mainly on roots, tubers, bulbs and other material within a short radius of its hole.

to vibration, and so the sound of thundering paws is unlikely to go unnoticed. Other wolves try to conceal themselves by stalking as low to the ground as possible and then, when close enough, taking a flying leap. This is a slightly more successful strategy, particularly for the older, more experienced wolves. For the less experienced, a bruised nose is a more likely outcome. But most of the time, the mole rat has already retreated into its hole, and so the challenge is how to get it out.

A wolf's first response to a mole rat that pops down is often to blow into its hole. The theory is that this scares the rat into moving, and from the underground noise the wolf can pinpoint where it is and so where to start

▲ **Rat-stalking.** In a characteristic rodent-stalking pose, an Ethiopian wolf sneaks slowly up on an unaware mole rat busy cropping plants around its hole. If the mole rat senses it and pops down the hole, the wolf's best strategy is to stand over the hole and wait for the mole rat to emerge again. Meanwhile it listens for underground movement.

digging. Youngsters can be comically bad at it. They're quite capable of digging themselves into huge holes in pursuit of a mole rat. They don't seem to know when to stop, and even if they are eventually successful, they may well have expended more energy digging than they get from eating their prey.

The most successful strategy, and the one employed by the most experienced wolves, is patience. A wolf will stand over a hole and wait for the mole rat to pop up again. Until that happens, the only movement the hunter makes is to occasionally cock its head from side to side and swivel its ears in an effort to keep tabs on any underground movement. If the rodent emerges – and it doesn't always – the wolf will pounce and grab it with its jaws.

THE SOCIAL PREDATOR AND THE SOCIAL PREY

An adult bull buffalo may stand 1.7 metres (5.5 feet) at the shoulder, weigh up to 900kg (1984 pounds) and have a set of horns 80cm (32 inches) across. Female buffaloes are smaller but not by much. And what they lose in size they make up for in numbers. Herds can be several hundred strong – a force to be reckoned with – and so for these plains herbivores, having nowhere to hide is rarely a concern. If threatened they can also be very aggressive. Indeed, buffaloes kill many people each year. The only predators buffaloes really need worry about are lions.

Single lions do, occasionally, kill a buffalo, but with prey this big and dangerous it takes teamwork – and experience. Lions that hunt buffaloes tend

▼ **Uncertain outcome.** Lions attempt to catch a newborn calf as a herd of buffaloes come to drink, but they find themselves confronted by the mother and a charging bull. The young lions in the pride watch as the lionesses try to defend themselves – and then flee.

► (next page) A buffalo battering.
A lioness that fails to flee fast enough is attacked by a bull and narrowly escapes being trampled to death. Buffaloes often try to help other herd members, and the larger the herd, the better the defence. Stragglers are the ones that are caught.

to make a habit of it. Lionesses typically do all the hunting for the pride, but with such a formidable quarry, male lions are often conscripted to add muscle. For lions, the challenge is to try to pick off an individual from the herd. So the ideal is to surprise the buffalo by using cover to get close. Lions are slower and have less stamina than buffaloes, which are capable of running at 60kph (37mph), and so the chase distance is important. But that's assuming the buffaloes decide to run in the first place. With safety in numbers they're just as likely to face down the lions and go on the offensive.

If the lions do manage to jump an individual and bring it down, they may suddenly find themselves dealing with a dozen or more angry members of the herd. Unlike other ungulates, such as wildebeest or zebra, buffaloes regularly go to the aid of a herd member in distress, and so the tables could easily be turned. Buffaloes do kill lions.

The bigger the herd, the less risk there is of lion predation. In fact, being part of a large herd is a better defence strategy than running away. So for lions, the best bet is to target one of the young bulls, which tend to hang out in small bachelor groups. If the bulls graze near cover, such as alongside a river where the forage might be better, they become much more vulnerable to a surprise attack.

Other good lion ambush points are waterholes, which buffaloes must visit every day. Using cover, they will wait until the herd has passed and try to pick out a straggler – or perhaps an old or weak animal. But buffaloes get wise to these dangers. If they have a choice, the herd will avoid waterholes where they have experienced a lion attack in the past few months. They will also visit waterholes at times when lions are unlikely to be active.

During the dry season in Zambia's Luangwa National Park, the temperature regularly hits 45°C (113°F). The heat is so energy-sapping for large predators such as lions that by 8am most are sleeping in the shade. For a subadult buffalo bull on its own, this is obviously a good time to visit a waterhole. But nature and behaviour can always be relied on to surprise – as one of *The Hunt*'s shoots was to reveal.

One morning, the team spotted a young bull walking across the wide open plain on his way to a spring. He passed a small group of lions sleeping in the shade of a tree. If he was aware of them, he didn't show it. But the lions noticed him. Though the temperature was already into the 40s, some of them clearly thought this was an opportunity too good to ignore.

◀ **(top) Heat of the moment.** In the heat of the day, a male buffalo walks past three young lions resting in the shade. He fails to spot them, but they see him.

◀ **(bottom) Going for the kill.** Working together, all three lions pull down the buffalo. One attacks at the rear and one attempts to suffocate and subdue it by biting its nose. The end seems certain.

▼ **The resurrection.** Seemingly exhausted by the heat and their exertions, the lions leave the body and collapse back in the shade. Suddenly, battered and bloody, the buffalo rises and walks back to the herd.

One lioness and two subadult males set off after the buffalo. Approaching silently from behind, one jumped onto its hindquarters. The surprised buffalo wheeled around on his attacker, then ran. But the three lions cut him off, and while one distracted him from the front, the other two attacked from the rear. The buffalo twisted and turned in an effort to hook the lions with his horns, but the cats were too quick. After 15 minutes in the sun, the lions finally got the buffalo on the ground. While one tried to suffocate him with a muzzle-hold, the other two attempted to cut through its tough hide. To our experienced cameraman, it looked all over.

Then something extraordinary happened. Weakened and exhausted by the heat and their exertions, the three lions abandoned their prey and ambled back to the shade. Perhaps they, too, thought the buffalo was as good as dead. But a few minutes later, battered and a bit bloody, the buffalo got to his feet and continued across the plain – watched by the lions. He was still alive two days later, testament to the ultimate unpredictability of the relationship between predators and prey.

SUN, SAND AND SCAVENGERS

The problem of having nowhere to hide is particularly acute in deserts. Temperatures in the Namib – one of the oldest and largest deserts in the world – can hit 60°C (140°F). Survival here depends on finding a way to escape the punishing heat. For most animals that means being active at night and hiding under the sand during the day. Hotrod ants, however, do the opposite, using heat to their advantage.

These are the most heat-tolerant ants in the world, able to cope with temperatures 10 degrees higher than any other ant species. Amazingly, their activity actually peaks at midday, when surface temperatures can exceed 70°C (158°F). The reason they do this is simple: they prey on dead and dying insects that have succumbed to the heat. The adaptation that enables such a lifestyle is extra-long legs. These keep their bodies further away from the scorching sand – surprisingly the air is 10 degrees cooler just 4mm off the ground, the height of the ants – and also allow them to run very fast, which reduces the amount of time each foot spends in contact with the burning ground.

Though adapted to the heat, hotrod ants still run a thermal tightrope. A hunting trip can last up to 30 minutes and take them 50 metres (164 feet) from the safety of their burrows. Misjudge the heat, and they could easily end up like the sun-scorched victims they are hunting for. If ground temperatures get dangerously high, they are forced to return to their underground nests or climb strands of grass to get to cooler air.

▲ **Desert death gives life to a hotrod.** Out on the Namib Desert dunes at the hottest time of the day, a hotrod ant investigates potential food – a larger species of ant that has succumbed to the heat. A hotrod ant has extra-long legs to keep its body high off the hot sand, and when standing, it keeps at least one leg off the burning sand grains.

▶ **(top) The trap.** A mat of sand and sticky silk lies over the lair of the spoor spider (so called because the mat forms a slight depression that resembles a spoor, or footprint) – a trap for unwary insects.

(centre) The strike. An ant is seized by its legs, bitten to immobilize it, and pulled down into the spider's burrow.

(bottom) The sit-and-wait strategy. The 10mm spider waits for prey directly under the mat. In the heat of the day, it may keep cool further down its burrow, but with a silk signal thread attached to the mat alerting it to the movement of potential prey, usually ants or beetles.

Being able to hunt in the middle of the day gives hotrod ants a big advantage: there's almost no competition for food. But there is at least one other predator, the spoor spider, which is also active in the heat of the day.

This spider copes with the blistering temperatures by weaving a cloak of sand grains and silk, under which it hides in a shallow pit, constructing a 10cm-long (4-inch), vertical, silk-lined tunnel as a retreat. It lines the edges of the mat with a network of sticky threads, and attaches signal lines to the underside, which it keeps hold of, to alert it to the vibration of an ant touching the trap. The moment it senses an ant, the spoor spider jumps out, grabs it by the legs, bites it and drags it down into its tunnel. If the temperature outside is just too hot, the spider will beat the ant at its own game by letting the sun finish it off before pulling it underground.

THE FINE ART OF TEAMWORK

The centrepiece of Namibia's Etosha National Park is a 120km-long (75-mile) saltpan, so large and flat it can be seen from space. It looks like the surface of an alien planet, and in October, the temperatures can hit 48°C (118°F). But what makes it possible for a surprising number of large animals – springbok, oryx, wildebeest, zebras and giraffes – to survive in the park are the numerous waterholes, which are where the animals regularly congregate. You might think that hunting at these spots would be easy for the lions. It isn't. With so little cover around the springs, it's almost impossible for the lions to sneak up on their prey unnoticed, especially in daylight.

To survive in this open landscape, the lions have had to become much more cooperative, which in turn has meant large pride sizes – some of the biggest in Africa. The main function of a lion pride is not to hunt but to raise and protect the young – from other lions. Indeed, the success rate of one lioness, or two hunting together, is almost the same as, say, six – except, that is, in Etosha.

In the 1990s, a long-term study of the Etosha lions showed a remarkable degree of cooperation, particularly when hunting springbok. To catch one requires at least half a dozen females, with each lioness taking on a specific role. Lighter females approach the target from the left and right. The job of these 'wings' is to drive the prey to the 'centres', which are generally heavier individuals. The hunts tend to be more successful if the same individuals

▼ **Hunters lined up.** As a storm rolls in, three lionesses in a large pride stand in anticipation of the start of a hunt. With little cover in the Etosha saltpan, the pride may use the noise and wind of a storm to mask the launch of their attack on an animal coming to drink at a waterhole, each lioness taking a specific role in the hunt.

play the same roles. So learning how to play a specific role works to the benefit of the whole team and enables the lions to survive in such a harsh, open habitat.

The lions also take advantage of ephemeral conditions. When the dry season finally breaks in December, storms provide the perfect cover for a hunt. Wind blowing through the low-level vegetation masks the sound and scent of stalking lions, and strong gusts pick up dust, reducing visibility that has already been compromised by the dark storm clouds. With their prey's senses dulled, the lions find it much easier to get close enough to launch an attack – and in these conditions they can be successful with far fewer members of the team. It's just one other strategy to get around the problem of having nowhere to hide.

CHAPTER FOUR

RACE AGAINST TIME

THE DYNAMIC BORDER BETWEEN LAND AND SEA – THE COAST – is one of nature's most demanding habitats. The animals that try to make a living here are under enormous pressure. On windswept coasts, powerful waves constantly smash against the rocks and cover the inhabitants with salty spray. Sandy and muddy shores are transformed every six hours by the daily rhythm of the tides. Yet for the predators and their prey, a coast can provide rich pickings. But the ever-changing nature of a coast means the best opportunities are time limited. So the challenge for both predators and prey is being there at just the right time.

▶ **Coastal survivor.** A marine otter, no bigger than a domestic cat, survives the pounding waves to bring an octopus to shore on the rocky coast of Chile.

◀ **(previous page) Racing for salmon.** A brown bear charges into the surf to catch one of the salmon gathered in the shallows off the coast of Alaska.

IN THE GRIP OF THE TIDES

If you really want to experience the true power of the tides, risk a walk into the middle of one of the world's great mudflats. These vast, featureless wildernesses may seem flat and boring, but they are the very best places to experience wide skies and teeming, undisturbed nature. The Wash on the east coast of England is one such place. A walk into the heart of this mudflat requires perfect timing. Start just after high tide and follow the retreating water. Within an hour or so, you enter a world without trees or houses, a world of wide skies and thousands and thousands of birds – waders and pink-footed geese mostly – that come here to rest and refuel.

The retreating tide opens the door on a rich larder. The mud that sticks to your boots is full of invertebrate prey that attracts vast numbers of wading birds. Around the world, a range of waders have developed their own ways of catching different prey. North America's long-billed curlew has the longest bill of any wading bird. It can reach more than 20cm (8 inches) into the mud and is perfectly designed to probe for shrimps and crabs. Godwits are smaller waders with long, slightly upturned bills that are extremely sensitive to vibration and feel for prey hidden in the mud. The diminutive plovers are search-and-run predators that snatch prey from the surface of the mud. Like many other shorebirds, they also tremble their feet to bring worms to the surface.

◄ **Mud larder.** The retreating tide off the coast of East Anglia reveals the great mudflats of the Wash, England's largest coastal bay. Within the mud are millions of tiny creatures that make the Wash an internationally important food source for waders and wildfowl.

Watching waders feeding and listening to the calls of passing geese can be hypnotic. But you need to make sure you turn back for the safety of the sea wall at just the right time. The returning tide comes racing in at a frightening speed, and unless you get a head start, there is a real risk you will get caught out on the mud and could even drown. For the pink-footed geese roosting out on the mudflats of the Wash or feeding on eelgrass, there is alternative sanctuary on the surrounding farmland. As the tide rushes in, vast v-shaped skeins of geese pass overhead, calling as they go. But for the waders, only the mudflats provide the food they need, and their larder is about to close for at least six hours.

▲ **Waders on the Wash.** A huge flock of roosting knots at high tide on England's Wash. When the tide goes out, tens of thousands of the knots move back onto the mudflats to feed, probing mainly for small, thin-shelled molluscs such as Baltic tellins, cockles and molluscs, swallowing them whole.

Desperate to grab every last feeding opportunity, flocks of waders chase the front of the returning tide, forming a rolling wave of hungry birds. As the area of exposed mudflats continues to shrink, the enormous numbers of waders that were spread wide across the whole of the mudflats of the Wash are concentrated into denser and denser groups. By the time you have made it back safely to the sea wall, tens of thousands of waders are swirling together in spectacular flocks that, at a distance, look like twisting clouds of smoke. This provides one of the British countryside's greatest natural spectacles.

PREDATORS ON THE WING

By the autumn, the number of waders using British mudflats reaches a peak. On the Wash, the vast majority are knots, up to 100,000 of them, escaping the bitter winter of their breeding grounds in Arctic Canada and Greenland. Peregrine falcons also exploit this bounty. Having spent the summer breeding on the uplands, they return to the coast to hunt through the winter, depending on speed rather than surprise to catch their prey on the wing. The knots defend themselves by forming ever-denser flocks, twisting and turning to confuse their attackers. Studies have shown that the bigger the flock, the less successful the peregrine, and so for an individual knot, there is obviously some safety in numbers.

Sparrowhawks, which rely on surprise rather than speed, are also drawn to estuaries in winter. These hawks have rounded wings that make them perfectly adapted to the agile work of hunting among trees. So when they come to the coast, they prefer to wait at the edge of any woodland bordering an estuary. Their favourite prey is redshank – a thrush-sized wader with red legs and a distinctive piping call. Redshanks search out sand-hoppers, which occur in their greatest density among the salt marshes that form a border between the mudflats and the woodlands where the sparrowhawks are waiting. The redshanks seem to sense the danger of ambush, but when the rising tide drives them off the mudflats, they have no option but to resort to the salt marshes. Once a sparrowhawk finds a close-enough perch, it will dash out and try to surprise a redshank.

Like knots, redshanks know that there is safety in numbers. Larger flocks confuse the peregrines, while many eyes provide greater vigilance and make it harder for a sparrowhawk to succeed through surprise. Redshanks have also learned to react differently to the two birds of prey. When threatened by a peregrine, which relies on speed and a persistent pursuit, they stay firmly on the ground. When they spot a sparrowhawk, the redshanks take off straight away. They obviously know this agile hawk is much more successful taking its prey on the ground.

▶ **Peregrine punch.** In a high-speed dive, a peregrine falcon has punched a knot out of the sky and now grabs it as it falls. By flying low over the flock, it has deliberately made the knots fly up from their roost. Though flocking together lessens the odds of being caught, peregrines are skilled at spotting which bird to pick out of the crowd.

◄ Marching out. As the tide falls, hundreds of soldier crabs emerge. Their defence against birds is sheer numbers, the ability to run fast forward (most other crabs only run sideways) and a speedy corkscrew exit into the sand.

► Sand sucking. At high speed, a soldier crab stuffs sand and water into its mouth, eating the tiny organisms and detritus that float into the water. It may shovel sand for several hours until the sea floods in.

OUT ON THE SAND

The daily tidal cycle that dominates the lives of predators and their prey on mudflats and estuaries has just as powerful an influence on those of sandy coastlines. Some of the world's greatest tidal ranges occur on the wild beaches of Western Australia, where in spring there is a difference of more than 10 metres (33 feet) between high and low tide. Though the endless expanse of sand might seem lifeless, get down to sand level, and you realize there is prey in abundance.

As the tide starts to fall, hundreds of tiny holes appear in the soft sand at the top of the beach, each with a pair of twitching eye stalks. Sand bubbler crabs, each no bigger than a pea, are waiting to emerge to sift through the exposed sand for organic matter. But lots of predators are waiting for them – waders, gulls and even kingfishers – and so only when all the beach is exposed will they emerge, en masse, swamping their predators with sheer numbers. Even then they are cautious, foraging in bursts and never far from their burrows.

As the tide falls farther down the beach, soldier crabs emerge to feed in 'armies' hundreds strong and swarm across the beach. The dark trail of freshly grazed sand they leave behind is followed by a caravan of predators. Herons and egrets join waders and kingfishers. But the ability of soldier crabs to walk forward as well as sideways makes them more agile than other crabs. Also, when danger threatens, the armies break up into smaller groups, which move in different directions to confuse the predators. If the threat is immediate, the crabs perform a rapid disappearing act, corkscrewing down into the sand.

The art of sand-shovelling

In Western Australia, the tiny sand bubbler crabs pop out of their burrows and start work the moment the tide goes out on their beach home. They can only sieve the sand when it's wet and so feed at breakneck speed before it dries out. Working systematically across the top layers of the beach, a crab passes the grains of sand through its mouth, filtering out detritus and microscopic animals, and then discards them in the form of small balls ('bubbles'). Timelapse shows how much ground a sand bubbler can cover in a few minutes. As it scours the sand for food, it leaves behind lines of small balls of processed sand – many balls a minute. After a while, a series of these lines radiates out from each burrow like spokes in a wheel. Soon the top of the beach is covered with hundreds of these wheels, perfectly spaced – no crab trespasses on a neighbour's patch. But this sand art doesn't just mark out territory. It seems that, with predatory birds ever present, the neatly laid lines of bubbles guide them quickly back to their burrows and safety.

▲ **Sand-bubbling.** As fast as sand is scooped into the mouth area (to be scoured of tiny organisms and organic matter by specialized mouthparts), sand is rolled out in a wet ball and shovelled back between the crab's legs.

▶ **1–4 Twelve-minute sand work.** A sand bubbler can work so fast that it may expel more than ten balls a minute. It moves in a circular or semi-circular radius of its burrow, never too far from its escape hatch, the route marked by the last row of sand balls.

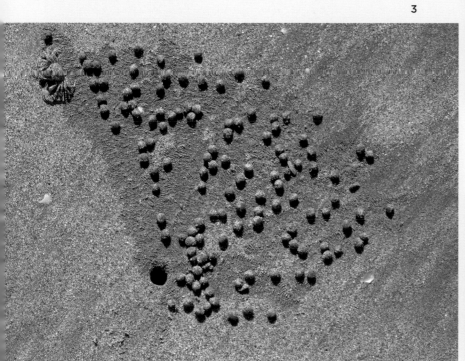

HUNTING ON THE ROCKS

Australia's tidal beaches are home not only to some very clever escape artists but also to equally ingenious predators. Among the most extraordinary is the recently discovered *Abdopus* octopus of northern Australia, Indonesia and the Philippines. Ranging in size from a golf ball to a tennis ball, this predator lives between the tides on reef flats, hunting in the shallow pools that are left when the tide retreats. Camouflage – the ability to change colour and shape – is an important part of both its hunting strategy and its escape from its own predators, as it is with most octopuses. But when a rockpool runs out of suitable prey, the *Abdopus* octopus does something surprising. Emerging into the glare of the tropical sun, it climbs out of the pool and 'walks' across the drying surface of the reef until it finds a suitable new rockpool with a fresh supply of prey.

Rocky-shore predators and their prey must not only be able to adapt to the tidal cycle but also the destructive force of breaking waves. Some of the

▲ **Time to move pools.** Having eaten what's in one rockpool, an *Abdopus* octopus sets off to find a new one with fresh prey trapped by the receding tide. As it 'walks', it adjusts its skin colour and pattern to match the rocks.

▲ **Little otter, big eater.** A marine otter feasts on a crab it has extracted from a boulder bed below Chile's cliffs. It doesn't have any blubber to protect it against the cold Pacific water, and so its dives have to be fast and efficient.

stormiest rocky shores occur along the Pacific coast of South America. This is where the marine otter survives. The smallest of all marine mammals – and much smaller than a European otter – its size and reduced body fat mean it quickly loses body heat when it dives. So it has to ensure it spends just enough time in the cold water to catch prey but not so long that it gets chilled.

Most of the otter's favourite prey – in particular crabs – are found on the rocky seabed. Though it appears to be able to hold its breath for a little more than a minute, its hunting technique is to go for short, fast-paced dives, and it can swim up to 2km (1.2 miles) in about 20 minutes. It also tends to favour more sheltered bays, to escape the worst of the destructive force of breaking waves. Until recently nobody knew exactly how the otters hunted beneath the waves. But as *The Hunt* team discovered (see page 282), it may be that the need for manoeuvrability is the main reason these otters are so small, allowing them to squeeze into the cracks between boulders on the seabed.

◀ **Axe and anvil.** A young long-tailed macaque concentrates as it uses an 'axe' to crack open a cockle balanced on an anvil it has chosen for shell work. Its troop are regular beachcombers along a coastline in southern Thailand. They live mainly on seafood and work in the hours of low tide, when the food is accessible.

▶ **Hammer precision.** Using a specially chosen stone as an axe-hammer, an adult macaque expertly knocks rock oysters off the rocks at low tide.

WHEN TOOLS ARE NEEDED

Animals that live permanently on wave-battered rocky shores need protection. Barnacles, limpets and oysters have fortress-like shells that they clamp onto the rocks. Crabs and lobsters have a coat of armour. Predators that want to eat such prey therefore need to have special adaptations, such as an oystercatcher's chisel-like bill, or be very strong or very clever.

Among the clever ones are the crows and the gulls that learn to fly up high and drop mussels onto rocks to shatter them. But the cleverest of all the rocky-coast predators have to be the tool-using long-tailed macaques living on the coast of Thailand. They choose different tools to deal with different shellfish. To knock oysters off the rocks, stone axes are used. The largest recorded axe weighed 1.7kg (3.75 pounds) – a heavy tool for an animal that averages just 5kg (11 pounds). The size of axe chosen depends on the size and location of the mollusc.

Sometimes, they also use hand-sized auger shells as pickaxes to prise oysters off rocks. And for a range of other shellfish, such as cockles and topshells, hammers and anvils are used. In fact, these remarkable primates search for shellfish along the coast much like human foragers and are the only animals other than humans known to use stone tools to kill and prepare animal prey.

GRASS FOR SOME, SALMON FOR OTHERS

The daily rhythm of the tides is not the only factor that dominates the lives of coastal predators and their prey. The ocean itself can be seasonal. This is driven by the warming influence of the sun, which fuels the cycle of phytoplankton blooming, which in turn fuels the whole marine ecosystem. Many of the best feeding opportunities for coastal predators are tied to this seasonal ocean cycle. And to exploit the bounty, they need to arrive at the coast at just the right time.

▲ **Nothing to eat but grass.** A mother brown bear and her cub make the most of the lush grass of an Alaskan coastal meadow – their staple diet from spring to summer, until the salmon start to move into the estuaries. Even then they have to keep a distance from male bears, which are capable of killing cubs.

One of the most spectacular coastal gatherings occurs every summer on the shores of the Katmai National Park in Alaska – a wilderness area the size of Wales (more than 1.6 million hectares/4 million acres). Katmai is home to the highest density of brown bears on the planet, more than 2000 of them. It's perfect bear country. Snow-capped volcanoes provide a place for winter dens. Lakes and rivers that crisscross the wilderness are full of fish in the summer, and lower down, the woodlands are a rich source of berries in the autumn. But the real bounty, and the reason the Alaskan brown bears are among the largest in North America, is found at the coast.

Each summer more than a million sockeye salmon return to the rivers of Katmai to breed. They have spent the past two or three years feeding up at sea. Now they need to swim upriver, a journey of up to 48km (30 miles), leaping waterfalls and struggling against the current to reach the gravel beds of the headwaters, where they lay their eggs. But before they can even start, the salmon have to prepare their metabolism for the change from salt water to fresh. So for about six weeks each summer, a million salmon wait in the shallow waters along the Katmai coast until their bodies are ready to start the long journey upriver. This is the opportunity the bears have been waiting for.

All Katmai's Alaskan brown bears spend the winter holed up beneath the snow in their high-mountain dens. After six months with nothing to eat, they are starving. The females with newborn cubs are the last to emerge from their dens, at the end of April, and are particularly desperate for food. But they face a long descent, just as the sun's new warmth is setting off avalanches. And when they reach the coast, the salmon are still at sea. The bears therefore have no option but to graze on grass in the coastal meadows – a far-from-satisfying food. This is a tense time. The meadows are packed with hungry bears, and for the next few months the mothers have to be wary of males that would like nothing more than to supplement their vegetarian diet with a bear cub.

By mid-July, the wait is almost over, and the bears start to move down to the beach. Though the salmon are still not in the shallows, the bears seem to sense that the dinner gong is about to go. All along the tideline, expectant bears are looking out to sea, standing up on their back legs, almost as if willing the salmon to appear. They are not alone. The heads of bobbing harbour, or common, seals appear in the waves, and ravens fly in from the mountains. Even the occasional shy wolf skulks along the beach, wary of the bears. Then suddenly, with the first silver flash of a salmon leaping out of the surf, the feast is on. For the next six weeks, the Katmai coast is alive with predatory action.

This national park is so remote, with almost no roads and no hunting, that the predators have little fear of humans. You can sit on the beach within just a few metres of a huge bear or a remarkably tame wolf and enjoy watching their extraordinary hunting skills and strategy. First to the table are the large male bears, which can wade out into the big surf where the salmon make their entrance. But in the beginning, even the oldest bears seem to have forgotten

Beach watch. A brown bear and a wolf patrol an estuary beach in Katmai, Alaska, waiting for the migrating sockeye salmon to come close enough to catch. The wolf is also shadowing the fishing bears, waiting to grab any scraps.

their fishing techniques. Time after time, they throw themselves into the surf with giant splashes only to see the salmon slip out of their paws. Even when one finally grabs a fish, he may have his catch stolen with the swipe of a paw.

This battle of the giants in the surf is no place for the female bears and their cubs. They get their chance as the season progresses and the salmon start to gather in the estuaries. Undisturbed by aggressive males, the females hunt with a different technique. Rather than trying to pounce on leaping salmon, some use the calmer river conditions to search with their heads under the surface.

Snorkelling works well for single females but is not ideal for a mother with cubs in tow. These families wait for the tide to turn and the salmon to surge upstream. When the river is packed with salmon flapping and twisting in the shallower water, fishing becomes possible for a mother bear, who can keep an eye on her cubs on the bank.

The tidal surge of salmon is also the opportunity the wolves have been waiting for. Like the ravens, they are masters at stealing scraps off the bears. But in the shallows, they can now catch whole salmon for themselves. In fact, these highly intelligent hunters are far more successful at fishing than the bears.

By late July, the first salmon will be moving upriver to spawn. The bears will follow, gathering at waterfalls to snaffle leaping salmon. This bounty is vital to all these predators' survival. In fact Katmai's Alaskan brown bears get nearly 90 per cent of their annual food supplies in just six weeks of the salmon run, which continues through to October. By comparison, the grizzly bears that live higher up in the mountains and never come down to the coast have to survive mostly on berries and are smaller than their salmon-eating relatives.

LOVE ON THE BEACH

The arrival of salmon at the coast dominates the lives of many predators in coastal regions worldwide, but there is one fish that makes an even more dramatic coastal appearance. Each summer, millions of capelin journey to the coast of Newfoundland in the North Atlantic to spawn. What makes capelin remarkable is that they are one of only two fish species that emerge from the sea to lay their eggs on the beach. (The other is the grunion, which breeds in far smaller numbers along the coast of southern California and Baja California.) The capelin arrive off Newfoundland in such vast numbers that a whole range of different predators have become almost totally dependent on the energy they bring.

At spawning time, the waves breaking on many Newfoundland beaches are rolling cascades of capelin. These slender fish, no more than 25cm (10 inches) long, shimmer silvery-blue in the surf. Soon all along the beach

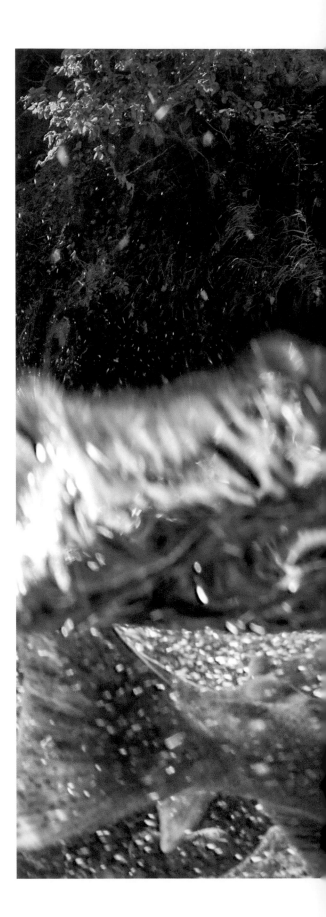

▶ **Salmon feast.** A female brown bear catches a salmon trying to swim upriver to its spawning grounds. Leaving coastal fishing to the larger males, young bears and females with cubs usually wait for the salmon to move up the rivers before starting to fish.

hundreds of thousands of silvery capelin are twisting and turning on the sand. This is a dance of love, with every female fish closely attended by two males. She will produce between 6000 and 14,500 eggs, which she leaves to incubate in the sand. Female capelins that make it off the beach will not return again that year. But males will hang around in the shallows and try their luck four or five times more. For a month or so, the beaches throng with writhing capelin – an irresistible draw for many predators. Flocks of gulls snatch fish from the breaking waves, and red foxes take so many that they have to cache them in the sand for later. Newfoundland's population

▼ **Unlucky lovers.** Millions of capelins lie dead or dying, having stranded themselves in their efforts to spawn in the sand and gravel on a Newfoundland beach. Awaiting their arrival along the coast are many predators, including larger fish and humpback whales.

of bald eagles, between 300 and 600 pairs and one of the largest in North America, is heavily dependent on these fish.

Nobody is sure why capelins leave their natural element to spawn on the beach. It may be that the warmth of the sand means their eggs hatch faster. Or it might be a way of reducing the pressure from their main predators in the ocean. Capelins play a key role in the marine food web, and large fish such as cod, herring and halibut await their return each summer. Many of Newfoundland's seabirds time their breeding so their chicks are hatching just as the capelin arrive to spawn. But by far the largest predators attracted by the massive shoals of fish are humpback whales. The world's greatest aggregation of feeding humpbacks comes to these waters every summer. Leaving their breeding grounds in the Caribbean, the whales journey up the east coast of North America, timing their arrival at peak capelin-spawning time.

The challenge for humpbacks is the fact that the capelins come into very shallow water to spawn. The whales' main hunting technique is lunge-feeding, when a humpback lunges up to engulf a large amount of fish-full water with its giant mouth and highly expandable throat. The fish are then sieved out with plates of baleen that run along each side of the mouth. With the longest flippers of any whale, humpbacks are particularly manoeuvrable and their lunges are high-speed – essential for catching fast-moving fish such as capelins. But this technique is also incredibly expensive energetically, and so it is only worthwhile targeting dense shoals of fish. When they are spawning, capelins tend to stick to the bottom, and so the whales need to get them up into the water column in shoals dense enough to be engulfed. Hungry fish such as cod seem to help with this, driving the smaller fish off the seabed. But in Newfoundland, the humpbacks use other techniques. Hunting right alongside the cliffs they try to corner the capelin. The manoeuvre nvolved are risky for 35-tonne whales, and they tend to choose smooth cliffs where there is less chance of injuring themselves.

Until recently lunge-feeding had only been seen at the surface, but using electronic tags on humpbacks off Newfoundland, scientists have discovered that they also lunge-feed at the depths where the capelins are concentrated. At night, they produce mega-clicks as they hunt, and it could be that, in the manner of dolphins, the whales are using their mega-clicks as sonar, scoping out the sea floor and scanning for shoals of fish. They might even be using them to concentrate the capelins and herd them up into the water column where they can lunge-feed easily.

THE BORDERLAND NURSERY

It is not only fish that risk predation when they come to the boundary between land and sea to breed. The ancestors of today's sea turtles were land-based reptiles, and the eggs of today's turtles still need to be laid on dry land. Most turtles tend to nest on remote islands where the range of predators is smaller, and they usually breed en masse as a way of diluting the pressure of predation. The tiny sandy Crab Island off Cape York Peninsula in Australia is one of just a few isolated breeding sites for endangered flat-backed turtles. When these metre-long (3-foot) turtles drag themselves out of the surf, you would have thought their rigid carapace would be defence enough. But waiting for them on the beach are saltwater crocodiles. These 6- or 7-metre (more than 20-foot) monsters have swum all the way up the coast to arrive at Crab Island just as the turtles arrive to breed. Saltwater crocodiles have been seen to pick up whole turtles, throw them into the air and then catch and crush them in their powerful jaws.

Even if the adult turtles manage to escape the gauntlet of crocodiles and lay their eggs, a range of other predators are waiting in the wings. The hatchling turtles tend to emerge under the cover of darkness, but even then there are hundreds of night herons picking them off the beach. Some pelicans even take beakfuls of sand and sieve out the hatchlings. And if the babies reach the sea, there are fish and sharks waiting to eat them.

One turtle that nests on mainland shores in great numbers is the olive ridley. Along the coast of Costa Rica, for instance, these turtles and their hatchlings have to deal with predators as large as jaguars, as cunning as coatis and raccoons, as agile as frigatebirds and as determined as ghost crabs and even ants. Their response is to concentrate their egg-laying into just a few nights. Tens of thousands of olive ridleys synchronize their nesting into a mass spectacle called an arribada. In some places, so many turtles crowd the sand that it would be possible to walk from one end of the beach to the other on turtle backs alone.

THE NEED FOR WARMTH

Many marine mammals also have to return to the coast to breed but for different reasons. All the pinnipeds – the seals, sealions and walruses – find it too energetically demanding to give birth and suckle their pups in the sea. In higher latitudes, where the ocean freezes over, many seals give birth on the ice. The

▲ **Arrival of the olive ridleys.** Thousands of female olive ridley turtles haul up at Ostional, Costa Rica, in November, at the beginning of the annual arribada – the mass beach-nesting event. Though some turtles are caught by jaguars and crocodiles, many more predators (including humans) are after their eggs and, later, the turtle hatchlings as they race to the sea.

risk in the Arctic, of course, is predation by polar bears, and so northern seals spend as little time as possible suckling their pups. In Antarctica, the danger is from leopard seals, which snatch other seals off ice floes, and pods of killer whales, which will try to wash them off in coordinated wave-making charges. Ice also offers little shelter from the wind and has a worrying tendency to break up. So where possible, marine mammals prefer to breed on dry land.

On the island of South Georgia in the South Atlantic, the beaches are so crowded with seals that it is hard to find any open space. Nearly 4 million

◀ **Desert dinner.** A brown hyena has its pick of the young seals trying to escape into the surf. It is one of a small population of hyenas that live on the Namibian coast, feeding almost exclusively on young Cape fur seals in the large breeding colonies and scavenging still-born pups and afterbirth. The coastal desert behind means few other competing predators reach this feeding ground.

▶ **Winners and losers.** A brown hyena carries a Cape fur seal pup up the beach. It is watched by a black-backed jackal, which will follow the hyena into the dunes, hoping to scavenge any remains of the meal.

Antarctic fur seals – more than 90 per cent of the total population – pack the beaches in summer, and along one 3km-long (1.9-mile) beach, more than 5000 southern elephant seals form a wall of blubber.

It is not the threat of predation that encourages such impressive aggregations, though. The main reason seems to be social advantage. The males of all these seals have many partners and fight each other for dominance and access to females. The pups feed entirely on their mothers' milk, and so the males have nothing to do but fight and mate. The females benefit by getting to mate with the fittest partners.

Though lots of scavengers are drawn to the coasts where seals are breeding, in particular birds, land-based predators are relatively rare. But on the coast of Namibia, bordering the great Namib Desert, there is a population of brown hyenas that preys and scavenges almost exclusively on Cape fur seal pups. Half of the 800–1200 Namibian population of this endangered hyena make a living on the coast, where pups are available year round, because of the long lactation period of the seals (approximately 11 months). And with no other predators in this tough environment, except for the smaller black-backed jackals, they have this concentrated coastal bounty virtually all to themselves.

IT CAME OUT OF THE SEA

For marine creatures to leave the sea to seek their prey on the land is almost unknown. But there is one highly organized predator that almost does that. Some killer whales that specialize in hunting marine mammals have learned that there can be rich pickings on the coast at seal-breeding time.

The Crozet Islands in the South Atlantic are one of the most remote places on Earth, but each year a pod of killer whales turns up just as the elephant seals are returning to breed. These deep-diving seals are well insulated with blubber, and when a killer whale grabs one in the kelp, the ocean turns red with their oxygen-rich blood. It seems that the same killer whales return each year, and similar behaviour has recently been observed in the Falkland Islands. It appears that the culture is passed down the generations.

Along the windswept coast of Argentina's Peninsula Valdes, small groups of South American sealions gather to breed. One group of killer whales has learned that, if they arrive at just the right time, they can steal the sealion pups right off the beach. But they have to concentrate most of their attacks into just a few weeks, from late March into April, so the sealion pups are just the right age. If the killers come too early, the pups are suckling their mothers higher up on the shingle beach. If they come too late, then the pups are no longer naïve and are far harder to snatch, having learned that great danger lurks in the breaking waves.

To have a better chance of stealing a pup, some killer whales that visit the sealion rookeries have learned a spectacular stranding technique. Swimming at speed, a whale launches itself through the breaking waves and onto the beach to grab an unsuspecting pup or one that doesn't move fast enough. The risk is being left high and dry on the beach, and quite often you will see a killer whale thrashing on the beach, trying to work its massive body back into the surf. Depending on the spot, they work either exclusively at low tide or at high tide and so obviously know the coast well. For a high-tide attack to have any chance of success, the killer whales also have to time it to when the tide is close to its highest, when they can strand closer to the pups without risking getting stuck. But there is also one channel that offers deeper water between two reefs for six

◀ **The attack channel.** A 5-metre (16-foot) killer whale (known as Jazmin) powers up from a deep-water channel that runs along the beach towards an unsuspecting pup (left) in the surf. She's well practised at catching sealion pups from this favourite spot.

hours around high tide, which extends the hunting window. So many stranding attacks are attempted here that it has been dubbed the attack channel.

This stranding behaviour is so difficult that, currently, only ten whales practice it. Four different groups visit the Peninsula Valdes sealion rookeries each year, and usually only one or two individuals in each group attempt stranding attacks. But if successful, they share the spoils with the pods. Different groups have different tactics. Some hunt on the rising tide so that, if they get stranded, they know they will be washed off again later. Others rarely risk coming right out of the water and prefer to try to snatch pups in the surf zone.

One of the most extraordinary parts of this story is the hunting school. Often a stranding expert will take a sealion pup out into deeper water. With its catch still alive, it will use its tail to flick the pup high into the air. This seems to be the start of a macabre game of cat-and-mouse where less experienced members of the pod learn the skill of handling pups at speed. It seems that the dangerous stranding technique requires great dexterity as well as perfect timing.

▲ **Beach exercise.** The female Jazmin attacks at high tide from the deep-water channel and just misses her target pup. Alongside her is her own calf, watching and learning. The sealion pups are also quick to learn, which reduces still further the window of opportunity for the hunter.

▶ **(top) Aim and charge.** A sealion pup scrambles up the beach just in time to avoid Jazmin's charge. The beach-stranding move is mirrored by her calf, who may take a few more years to learn the technique – young killer whales don't usually attempt beaching until about ten years old.

▶ **(bottom) Grab and exit.** Jazmin grabs a pup in the surf and starts to thrash back down the beach with her catch.

CHAPTER FIVE
IN THE GRIP OF THE SEASONS

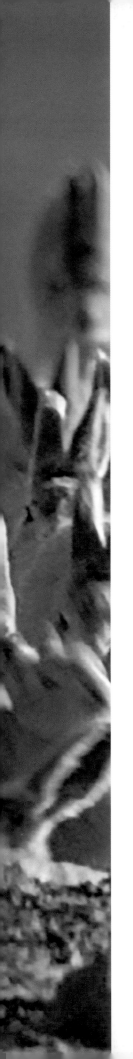

FOR PREDATORS AND PREY THAT LIVE IN THE POLAR REGIONS, the greatest challenge to survival is continual change. Nowhere else on the planet are the advancing seasons so keenly felt. Every winter the Antarctic continent effectively doubles in size as the sea freezes over. At the same time in the far north, two thirds of the Arctic Ocean sea ice melts away with the arrival of summer. On land, as the snow melts, the Arctic landscape is transformed from white to brown and green. How does an animal camouflaged to disappear against the snow deal with such dramatic change? For polar predators and their prey, the only way to survive is through adaptability.

▶ **The prey – Arctic hares.** Leverets race away from an Arctic wolf on Ellesmere Island, Canada. Once they are weaned, the young hares herd together, giving safety in numbers until the autumn snows arrive.

◀ **(previous page) The predator – Arctic fox.** Spring on Wrangel Island in the Russian far north brings the snow geese and provides Arctic foxes with eggs and then goslings. Here a fox carries a snow goose egg snatched from a nest, to be cached for later consumption.

KING OF THE ICE

The supreme predator of the Arctic is the polar bear. The world's largest land carnivore is perfectly adapted to the Arctic winter. While grizzly bears only survive northern winters by hiding away in dens beneath the snow, polar bears can stay out on the sea ice throughout the long dark nights. In fact, they are so well insulated that their main problem is overheating.

A polar bear's body temperature is almost identical to ours but the tips of its hair can be 75°C colder. Its insulating coat is not actually white but transparent. This allows light to penetrate and warm the bear's black skin. Much of the reflected infrared light is trapped by the hairs, which are also hollow – the air inside heats up to provide yet more warmth. A polar bear's total insulation is so effective that when you study one using a thermal camera there is no discernible loss of heat other than through its breath.

Its large size not only insulates it against the cold but is also essential for hunting. Without its strength and weight, the bear would never be able to break through the ice to get at its seal prey. But bulk comes at a cost. A polar bear uses twice the energy to do the same movements as almost any other mammal. The risk of overheating also means it has to move slowly, with limited bursts of speed. A polar bear's whole hunting lifestyle is a careful balance between energy gained from its prey and energy expended capturing it.

◀ **Diver, swimmer, ice-walker.** A polar bear emerges from the sea after an unsuccessful hunt. Though a swimming bear may sneak up on a seal resting on ice, once a seal is in the water, it's very hard to catch. A bear's insulating fur acts as a drysuit, and its body fat keeps it buoyant and enables it to remain in the water for many hours – even days.

THE BEST SEAL TO EAT

Ringed seals are a polar bear's favourite prey and usually make up more than 80 per cent of its diet. They are the most widely dispersed and numerous Arctic seal, with a population of more than 7 million, and the smallest, which means even small bears can handle them. But it's their lifestyle that makes them such important prey. While other Arctic seals live out in the moving world of pack ice, ringed seals stay closer to land on the fast sea ice – the permanent ice attached to the land that doesn't melt away even in summer.

Uniquely among Arctic seals, a ringed seal has five sharp claws at the ends of its flippers and can dig and maintain its own breathing holes (the other seals depend on holes that appear in moving sea ice). In winter and early spring, a seal often digs out a lair beneath the ice and frozen snowdrifts, where it rests up in the winter or nurses its pup in the spring, sheltered from polar winds and out of the view of patrolling polar bears.

During the long, dark Arctic winter, polar bears – mostly males but also non-breeding females – patrol the frozen ocean in an endless search for seals. The sun is just a distant memory or a subtle glow below the horizon. Howling gales bring wind-chill temperatures below −70°C (−94°F), and the sea ice looks lifeless. How do polar bears find their prey in this dark, featureless world? A fascinating experiment with a Labrador dog points to the answer. Researchers used a dead seal to give the dog the scent and let him loose on the sea ice in early spring. They were amazed not just how quickly he sniffed out the seal lairs beneath the ice but also just how many seals were hiding in them. And the density of breathing holes was far greater than they had expected. One reason for that is that each seal maintains up to five or six different holes to try to confuse its predators. But still, even in winter, there are obviously good numbers of seals to be caught if the bears can only break through to an occupied lair. The researchers were also surprised that their dog seemed to be able to smell a seal beneath the ice from more than a kilometre away. It would seem likely that a polar bear's sense of smell at least matches a dog's. After all, to a polar bear the sense of smell is everything.

▶ **Ringed seal – prey for all seasons.** A young ringed seal, smallest of all seals, basks on ice off Barren Island, Nunavut, Canada. Its camouflage-white puppy-fur has been replaced by a juvenile coat, dark above and silver underneath. The seal has been able to swim since it was a week old, and should it need to escape a polar bear, it will dive. At this age, its chances of being caught are not as high as they were before it was weaned.

MAKING A BREAKTHROUGH

A polar bear patrolling the sea ice walks with a slow, rolling gait, heading into the wind, or at least across it, increasing its chance of picking up a scent. Ringed seals hiding under the ice are sensitive to the slightest vibration, and ice transmits sound and vibrations effectively – microphones placed under the ice have picked up human footsteps 400 metres (1,300 feet) away. At any suggestion of a bear, a seal will slip into the ocean beneath. So polar bears have to proceed with extreme caution, placing each giant foot ever so gently.

Once it's on top of the lair, the polar bear slowly transfers most of its weight onto its back paws before suddenly rising up to its full height and pounding down on the sea ice. To have a good chance of catching the seal, the bear really needs to break through on its first pound. Usually only large, heavy and experienced males are successful the first time. One study in the Beaufort Sea, where ringed seals are numerous, recorded 556 attempts to break into lairs, of which only 46 were successful. For most bears, the success rate in winter is low, and non-breeding females often have to scavenge the remains of other bears' kills. It may well be that the challenge of breaking into ringed seals' winter lairs is the key driving force behind the great size of polar bears.

RELAXED SOUTHERNERS

In the Antarctic, the Weddell seal is in many ways the mirror of the ringed seal. It's far larger, but like the ringed seal, it stays throughout the winter, using its teeth to maintain breathing holes in the ice. But in Antarctica there are no land-based predators to threaten the Weddell seal when it gives birth in the spring (though in the sea, leopard seals and killer whales hunt Weddells).

Both harp seals and Weddell seals take six weeks to wean their pups, but the process is different. Arctic ringed seals are solitary, each pair maintaining a number of breathing holes to confuse their predators. Having no predatory pressure under the fast ice, Weddells use only a single hole, shared by a male and eight to ten females. Dive beneath the Antarctic ice and you hear a haunting chorus of calls as each male defends his harem and their precious breathing hole.

Unlike ringed seals, Weddells have nothing to fear out on the ice and spend many hours asleep in the open. And while all Arctic seal pups are white for camouflage, Antarctic seal pups are black or grey. The lack of any land predators has given Antarctica's seals and penguins a much more relaxed life on the ice.

▶ **(top) Hunting the naive.** In its white coat, a newborn ringed seal pup lies on the ice above its refuge, unaware of the polar bear approaching slowly and silently. The challenge for a bear is to stalk it without a sound or vibration – difficult for such a large animal.

▶ **(bottom) Small spring pickings.** Out in the open, a pup makes a small but easy meal for a bear. Had this one been just a few days older and able to swim, it could have escaped back down to the lair its mother sculpted for it under the ice and, if necessary, slipped out through the bolthole into the sea below. The challenge for the bear would have been to break through the ice cover in one massive hit before the pup had time to exit into the sea.

THE BEAR'S FOLLOWER

Polar bears are not the only predators that patrol the Arctic sea ice in winter. Arctic foxes are also year-round hunters. Their thick winter coats deliver both camouflage and insulation, and their small size, short muzzle, small ears and the fur between their toes all help to keep them warm. But winter life for an Arctic fox in the far north is hard.

These little foxes can't break into ringed seals' lairs. Instead, they have to rely on polar bears. Following behind at a safe distance, they scavenge from the polar bears' meals. It's only in the spring, when lemmings become accessible and the migratory birds return, that Arctic foxes can resume a lifestyle as true predators.

THE SPRING MENU

In the spring, the other seals return to the Arctic sea ice to breed, offering polar bears a wider diet but also fresh challenges. Bearded seals are a good deal larger than ringed seals. Rather than making lairs under the fast ice, they give birth at the edge of the firm ice where it starts to break up into pack ice. This moving world is harder for the polar bears to negotiate, and bearded seals are careful to ensure that their newborn pups are only vulnerable for a short time. The mother's milk is extremely rich, containing 50 per cent fat, and so pups can be weaned and take to the ocean at just six days old. And the large size of bearded seals usually means that only larger polar bears can deal with them.

Hooded and harp seals breed even farther out on the edge of the pack ice. Unlike the solitary ringed and bearded seals, both give birth in colonies often many hundreds strong. Though these colonies can be far from the fast ice, they are still a temptation for at least a few polar bears. But because breeding is synchronized and all the pups are born at about the same time, the bears are swamped with choice. Also, hooded seal milk is so rich that the pups are weaned faster than any other mammal – in just four days. So the breeding cycle is over in a few weeks, and the polar bears are forced to turn inland again in search of food.

◀ **Sneaking scraps.** The fox is likely to have followed the bear for a while, scavenging leftovers. Polar bears can be essential for the survival of Arctic foxes in winter, when no small prey is available. Some foxes will even follow bears throughout the winter, staying just out of reach of their providers, sometimes even nipping at their heels to distract them before dashing in and snatching a scrap.

WHEN THE SUN RETURNS

Seasonal change is rapid in the polar regions, and spring comes quickly to the Arctic. In Svalbard, just 960km (600 miles) south of the North Pole, the sun first rises on 14 February. After nine weeks, on 18 April, the sun shines continuously for 24 hours, and it will not set again until 24 August. The sun's warmth soon starts to melt the sea ice and fuels a green phytoplankton bloom, clouding the surface water. On the land, the disappearing snow reveals a brown and green tundra, carpeted in places with spring flowers. A world that was silent through the winter comes alive with the calls of thousands of birds returning from the south. Each year more than 150 species fly north to breed, attracted by the blooming of life in the ocean and the 24 hours of daylight that allow round-the-clock feeding.

All these summer visitors face the same problem. They have come from a world of trees and cover, but in the Arctic there is no place to hide. For the penguins in Antarctica this is not a problem, as there are no land predators to disturb their breeding – the main reason penguins have lost the ability to fly. But in the north, Arctic foxes specialize in stealing the eggs and chicks of breeding birds, while glaucous gulls, gyrfalcons and snowy owls also follow the avian vanguard north to prey on this winged bounty. And later in the summer, desperate polar bears will also raid the nesting colonies. All the birds that come to the Arctic to breed have had to find strategies to deal with these predators.

LIFE AT THE TOP

Most of the seabirds nest on steep rocky cliffs close to the coast. The cliff-face colonies may contain tens of thousands of breeding birds and form some of the Arctic's greatest natural spectacles. But they comprise just four species. The vast majority are guillemots – mainly Brünnich's but also common – which nest on the thinnest ledges. Their eggs have a special pointed oval shape that helps to ensure they don't roll over the edge. Two species of kittiwake – the black-legged and the red-legged – make nests on the wider ledges. Determined Arctic foxes are always trying to scale these avian tower blocks, but for the most part, the cliffs provide their residents with a good deal of security.

The most numerous seabird in the Arctic has another defence strategy. Little auks are barely larger than starlings, and this allows them to squeeze between the small rocks in the scree slopes beneath tall cliffs. Protected in

▲ **Arrival of the guillemots.** Thousands of common guillemots approach their colony on cliffs in the extreme north of Norway. The birds have wintered at sea, but in March they return to their nest sites and will breed as soon as the cliff ledges are free of snow.

these rocky chambers, the little auk's eggs and chicks are out of the Arctic fox's reach. The adults, though, still have to deal with aerial predators – glaucous gulls and gyrfalcons.

Gyrfalcons are the peregrines of the north. They come in two colour forms – grey and white – both perfect camouflage against the ice and snow. To help deal with attacks from the air, whether by gyrfalcons or gulls, little auks nest together in enormous numbers – the largest colonies in Greenland contain more than a million pairs. Adults returning from the sea with food gather in large flocks out at sea, flying in together as a swirling smoke of little black birds. When they

◄ **The cliff-breeders.** Brünnich's guillemots nesting on narrow ledges on a vertical cliff face on Svalbard. The choice of location keeps their eggs and young relatively safe from land predators but not from gulls. Egg-laying (one large egg per pair) is synchronized so that all the chicks hatch and are ready to make the leap and glide to the sea at roughly the same time, even though they can't yet fly, gambling on safety in numbers.

► **The boulder-breeders.** A pair of little auks in a coastal colony on Svalbard resting outside their nest chambers on a boulder scree slope below the cliffs. Relatively safe in holes and crevices under the rocks, little auk chicks have time to grow their feathers, and they don't leave their nests until they are able to fly out to sea.

reach the colony, they don't go straight to their nests. Instead they circle around the scree slope, forming a noisy doughnut of beating wings. For gyrfalcons – high-speed predators that catch their prey by surprise – entering a swirling swarm of little auks could be dangerous, so they prefer to go for more solitary prey.

Even though an Arctic fox cannot squeeze its way into a little auk's rocky nest chamber, this wily predator has developed a clever strategy. It slinks into the heart of the breeding colony and hides among the scree. Sooner or later a marauding glaucous gull or the lightning pass of a gyrfalcon will frighten the auks, which will take to the air. As they return to their nest chambers, the fox will leap out and suddenly grab one of them.

OUT IN THE OPEN

Other breeding birds – mostly waders, skuas and terns – have no option but to nest in the open on the ground. These exposed breeders have two totally different strategies to deal with Arctic predators. Skuas and Arctic terns go on the offensive to defend their nests, repeatedly stooping down and attacking any intruder. Anybody who has been dive-bombed by an angry skua will tell you what a terrifying experience it can be. Even polar bears have been driven away by angry Arctic terns, whose dagger-like beaks will draw blood.

Other ground-nesting birds make themselves as inconspicuous as possible. The masters of disguise are female eider ducks. The males are brightly coloured for courtship and display, but the females' plumage matches the tundra. On the rare occasion when they leave their eggs, they cover them in down plucked from their breasts. If a female is brooding and an Arctic fox passes by, she doesn't flee and risk the eggs cooling but keeps absolutely still and drops her heartbeat down to almost nothing for up to ten minutes, with barely a breath.

The three species of phalarope waders that come to the Arctic to breed are famous for their distinctive feeding technique – spinning like brightly coloured tops on the surface of the water with rapidly beating legs to disturb their prey. But their other claim to fame is role reversal when it comes to camouflage plumage. While male birds are usually the most brightly coloured of the sexes, using plumage to impress females and ward off rival males, and the females are drabber, making them less conspicuous on the nest, it is the female phalaropes that are brightly coloured. So it isn't surprising that male phalaropes do most of the incubation. But why?

All Arctic waders lay larger eggs than average, because their chicks need to do most of their growing inside the shell, so they are ready to leave the nest almost immediately after they hatch, reducing the time when they are most vulnerable to Arctic foxes. The relatively small phalaropes lay large eggs, and the energy required to do so puts a great strain on the females. So the exhausted mothers let the fathers take on the egg and chick care and often head south early, to fatten up and recuperate, ready to breed again next spring.

A LANDSCAPE TRANSFORMED

When the snow starts to melt in spring, the land in the high Arctic undergoes a much faster and more dramatic change than just the gradual disappearance of the sea ice. The vast expanse of tundra that was white all winter is transformed into a sculpted landscape of greens, browns and greys. From the air, you are struck by the beautifully regular patterns that the frost has stamped into the tundra and by the silvery veins of the rivers that flow free again. Vast herds of caribou trek up from the south, and the landscape comes alive with the calls of summer visitors.

▶ **Safety in numbers.** Adolescent Arctic hares, almost in their full adult white coats, feed in a group in summer on the tundra of Ellesmere Island in Arctic Canada. Herding together means more ears and eyes are alert to the ever-present danger of wolves.

For the few winter residents, the camouflage that protected them through the winter can become a hindrance. Of the two land-based predators, only the Arctic fox bothers to lose its white winter coat. The Arctic wolf remains white all year. The most likely reason is that these two predators have different ranges. While the fox is widely spread throughout the Arctic, the wolf is restricted to the high Arctic closer to the pole. The summer is so short at the most northerly latitudes that it may not be energetically worthwhile for the wolf to change colour, as there are only a few weeks of summer here when the tundra loses most of its snow.

▼ A mother's brood and a wolf's food.
An Arctic hare nurtures 11 leverets. Only
some of the adolescents are likely to be
her own in a gathering that is as much for
comfort as for suckling.

The wolf's favourite prey, the Arctic hare, doesn't bother to lose its winter coat either. Early in the summer, when the hares are giving birth, they shine out as beacons of white on the grey-green tundra. Much larger than their southern cousins, Arctic hares will stand straight up on their back legs to look out for predators. At any sign of danger, they flee, with a strange bouncing run, like small, white kangaroos. They are surprisingly fast, and wolves are forced to run flat out chasing them often for long distances.

Both predator and prey are equally matched in terms of speed, but when it comes to manoeuvrability, the hares win out. Only by watching from the air can you truly appreciate the Arctic hare's extraordinary ability to twist and turn within inches of the wolves' snapping jaws. A single wolf can rarely catch an adult hare, but wolves seldom hunt alone. If a pack is chasing, one wolf will often head out in another direction to cut off the hare's escape.

The hares, though, have one last line of defence. Towards the end of the summer the survivors will often gather together with their full-grown leverets in herds that may be tens or even hundreds strong. Many eyes watching provide safety in numbers.

THE BIGGEST COLONIES OF ALL

Among the most impressive of all the summer spectacles in the far north has to be the aggregations of snow geese that gather to nest each year. There are massive breeding colonies all around the Arctic, where you see nothing in every direction other than hundreds of thousands of white dots right to the horizon.

For the Arctic foxes that live in these areas, the annual return of the snow geese is the most important food event of the year. The large eggs and chicks make satisfying meals for foxes, but geese don't give up their offspring easily. Domestic geese are used as guard dogs for a reason, and their wild cousins are no less feisty. If a fox comes even close to their nest, the parent geese will run at it, hissing loudly and violently flapping their wings.

A single goose can see off a fox, but foxes have learned to work in pairs. While one distracts an angry goose, the other will nip in and steal an egg or chick. At the height of the season, the foxes are so successful that they cache their booty underground for the winter. To get protection from the foxes, some geese form an unlikely alliance with the Arctic's most impressive

avian predator, the snowy owl. The owl's favourite prey is the lemming, the high Arctic's only resident small mammal.

Lemmings have been found as far north as 82 degrees on Ellesmere Island, surviving the worst of winter in burrows under the snow. Throughout their range, lemmings change colour every summer, probably because the intense predatory pressure from snowy owls increases the need for camouflage. In a single season a pair of snowy owls may bring as many as 2500 lemmings to their chicks (see page 110).

These large and powerful predators will actively defend their eggs and chicks, swooping down and driving off any fox that comes within about 400 metres (1,300 feet) of their nest. Any geese incubating eggs in this safety zone are rarely disturbed by foxes. But later in the season, the snowy owls demand a protection price. When the goslings head off to the nearby lakes to feed, they make easy meals for the owls.

▲ **Snatch and run.** An Arctic fox races away with an egg grabbed from an unattended snow goose nest on Wrangel Island in the Russian Arctic. Too late, the parents launch an attack. The fox goes on to bury the egg in the tundra – its cold store – and returns to gather more eggs to cache to provide vitally important food for when times are hard.

▶ **Parents on guard.** Vigilant parents lead newly hatched goslings away from the nest to feeding grounds higher up on the tundra. Most of the snow goose colony's goslings hatch in a two-week period, and so the foxes have a short time of plenty and can only catch relatively few goslings.

TIME TO GET FAT

Spring and early summer is the one time of year that the Arctic's supreme predator has the odds stacked in its favour. It's when the polar bear's seal prey are giving birth to their pups. In just six weeks, from April to mid-May, most bears get 90 per cent of their food for the year.

In spring, the ringed seal pups hide in the lairs beneath the ice that their parents dug out in the winter. But as summer arrives, the hard covers of blown snow, which used to cover the lairs in winter, start to melt away. As the season progresses, the naive ringed seal pups come out on top of the sea ice, and the odds are even more in the polar bear's favour.

At the beginning of April, polar bear mothers and their newborn cubs start emerging from their winter dens. The mothers are starving, not having

▲ **Mother's meal.** Having just caught a ringed seal, a mother polar bear heads for a stable piece of ice where she can eat her catch. She is comparatively thin and is probably having a difficult time hunting with young cubs in tow, especially now the ice is melting.

eaten since they entered the den back in early November – and, in many cases, since the end of the previous summer. They are desperate to kill a seal to give them the energy to suckle their cubs. At the same time, they are wary of lone male bears that will kill their cubs, both for food and to bring them back into breeding condition. Indeed, the courting season coincides with the main seal-hunting season.

To avoid lone males, female polar bears with newborn cubs tend to stick to the fast ice close to the shore or in protected bays. This means ringed seals are their only potential prey, even though the hunting here tends to be more difficult. For these mothers, all spring hunting is made that much harder because her cubs do not understand the need to keep quiet. All they want to do is play. You can really see the mother's frustration when she turns towards her naughty cubs and grunts at them to keep up and keep quiet.

TWO WAYS TO CATCH A SEAL

At this time of year polar bears have two different hunting techniques – still-hunting and stalking. Of the two, they much prefer still-hunting because it is far less energetically demanding. It consists of the bear lying absolutely still on the edge of a ringed seal breathing hole, waiting patiently for a seal to return. The wait is usually less than an hour, but occasionally it can involve staying in one place for several hours before exploding into action and lunging into the hole.

Sometimes a bear will stay head down for some time after the initial lunge, blocking the hole with its hind legs in the air. This may be to cut off the light coming into the hole to attract a ringed seal by tricking it into thinking the hole is still covered with snow.

Stalking comes into its own a little later in the season, when the ringed seal pups are big enough to rest up on the ice, along with the moulting adults. The polar bear's challenge is to get close enough before the seal or pup realizes it is in danger and escapes back down the breathing hole to the ocean.

The bear has to approach downwind and move slowly and carefully. Seal pups don't have particularly good eyesight, but they are sensitive to even the smallest vibrations in the ice. The bear will need to get within about 20 metres (65 feet) before it is worth making an explosive run. It's a skilful strategy that takes practice.

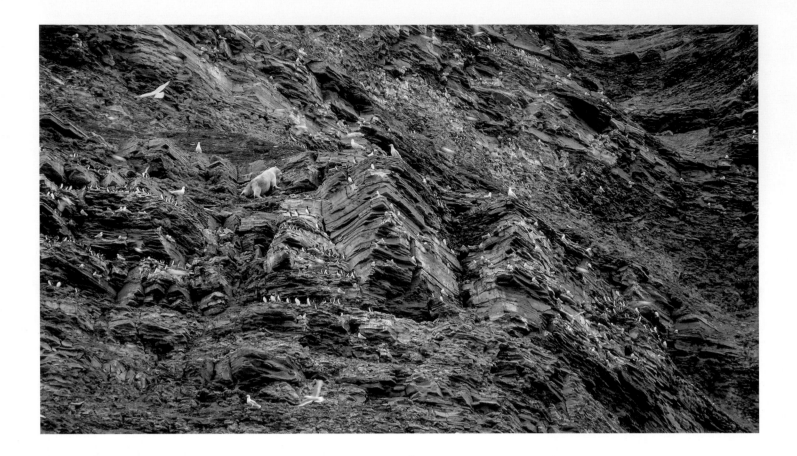

Desperation hunting

Polar bears are opportunists and will scavenge as well as hunt, and in summer, ground-nesting birds may offer easy pickings. Some bears have been known to spend many days on a single island, slowly working their way through all the nesting eiders, eating eggs or chicks or even raiding the nests of little auks, overturning boulders to get at them. Others are desperate enough to scale steep cliffs to try to get at guillemot nests. This is hard work for a bulky animal not designed for rock-climbing, and bears have been seen to fall hundreds of metres off a cliff, landing in the sea below.

Interestingly, polar bears have never been seen raiding snow goose colonies. With thousands of birds nesting close together, there should be rich pickings for bears when the chicks are small or the adults have moulted and can't fly. It seems the problem lies in the fact that even a flightless goose can move too quickly for a polar bear. The energy it would use to catch the goose would far outweigh the potential gain.

▲ **Scaling the heights.** A male bear takes a treacherous route up and across a crumbling cliff on a Svalbard island, with a drop to the sea below of more than 300 metres (985 feet). The bear's decision to make the dangerous climb will have been influenced by the lack of sea ice and the resulting difficulty of stalking seals.

▶ **1–4 Slim pickings.** For this male polar bear, the eggs and chicks he is stealing are a meal of desperation. Often a single bear will spend many days slowly working its way along the cliff to fill its belly, while glaucous gulls wait to pick off anything it leaves.

1

2

3

4

▲ **To chase or not.** An Arctic wolf ponders whether it is worth a sprinting attempt to catch the hare. Though Arctic hares can run surprisingly fast, they don't have the stamina of wolves, and on Ellesmere Island in the Canadian high Arctic, they are a staple wolf prey.

◄ **Sizing up.** A female with pups at the den takes an investigative walk past the photographer. On Ellesmere Island there are virtually no people and no hunting and so all the animals are unafraid of humans. Arctic wolves are a subspecies of the mainland grey wolf but smaller, with smaller ears and muzzle to help with heat conservation.

WHEN WOLVES NEED BIGGER PREY

Arctic wolves are smaller that their southern cousins, with small ears and short muzzles to help slow heat-loss in winter. Because prey can be thin in the high Arctic, wolf packs at these latitudes are rarely more than five or six strong. At the start of spring, when the pups are born, the pack may just be the mother and father. The pair can't go far from the den, which will be hidden beneath large boulders or in a hillside – often a traditional site that the wolves return to year after year. Once the pups are weaned, the parents start to bring them Arctic hares and lemmings that they have hunted relatively close by. Then once they are large enough and strong enough to leave the den, they form a family pack.

As summer progresses and Arctic hares and their leverets are no longer enough to feed the growing pups, the wolves go in search of larger prey. They can range over massive distances and have been known to travel 1000km (620 miles) in search of food. Their largest prey are muskoxen. These massive animals are designed to withstand the worst of the Arctic winter. While lemmings hide out in warm burrows beneath the snow and wolves can find shelter in their dens, muskoxen have no option but to stand out in the worst of the winter storms, relying on their large size and extremely thick and heavy coats to keep them warm. All through the winter, the muskoxen scratch away at the snow, searching for lichen to keep them going. These grazing spots become little oases in the snow, and ptarmigan will fly in to scavenge on scraps of lichen and roots the muskoxen dig up.

For even a large pack of Arctic wolves, it is a real challenge to take down an adult muskox. So they prefer to go for calves. The technique they use is to try to cause the herd to panic and run, in the hope a youngster will be left behind. Any straggler is attacked, but the battle is rarely over quickly. The other muskoxen will turn around and run at the wolves – and there can be 30–40 of them. They will also encircle their calves, all facing out towards the wolves – an invincible ring of adults. Only by separating a panicked calf from the herd do the wolves have a chance of success.

▶ **(top and bottom) The long pursuit.** A family pack of wolves on Ellesmere Island tries to run down a male muskox (calves are the easier prey) – a marathon that lasted more than an hour. The muskox, exhausted and cornered, eventually succumbed to the wolves, but it wasn't a foregone conclusion.

▼ **Family meeting.** Muskoxen gather directly above the wolf den, seemingly confident that the wolves here are preoccupied with pup-rearing.

WATER STALKING

By the end of May and into June, the Arctic sea ice has really started to break up, and most of the ringed seal pups have been weaned and have disappeared away to sea. Times have become increasingly hard for polar bears, and they have to resort to yet another hunting technique. The seals that remain haul out to rest on floating ice floes.

With its firm hunting ground melted away, a polar bear has no option but to swim. But it doesn't have the scientific name *Ursus maritimus* (sea bear) for nothing. Polar bears have been seen out at sea as far as 160km (100 miles) from the nearest land. A long neck allows a bear to keep its head above water, and its huge feet (the largest in the bear world) provide perfect paddles for swimming. But getting close to wary, hauled-out seals requires a special stalking technique.

Earlier in the summer the surface of the sea ice became increasingly pockmarked with melt holes, which some polar bears have learnt to use for an ingenious disappearing act. As soon as a bear has stalked over the ice to about 100 metres (330 feet) from a hauled-out seal, it knows it will soon

Silent hunter. A polar bear swims in the channels between ice floes looking for seals. Once a seal is in sight, it may approach so low in the water that only its nose is visible.

be spotted. So it will slip out of sight down a melt hole and swim towards the position where it thinks the seal is. But once under water, a bear can easily get disorientated. What often follows resembles a game – with the bear repeatedly popping up from under the ice to spy on its prey. Often it has its bearings all wrong and will emerge from a melt hole farther from the seal.

Later in the summer, as the melt continues, the sea ice is no longer a pockmarked solid sheet but a moving jigsaw of broken chunks. This is when true aquatic stalking comes into play. First the bear has to make sure as little of it as possible is visible at the surface. It swims low in the water with just the tip of its long nose breaking the surface to breathe. Once it has spotted a hauled-out seal – almost always a bearded seal – it tries to hide behind ice floes as it makes its approach.

Sometimes the melting process leaves little valleys on top of the ice. Keeping its body as flat to the ice as it can, a polar bear can use these watery tunnels to creep ever closer. When the seal is just one ice floe away, the bear will often swim right under it before suddenly bursting out of the water to grab it.

Bear alert

In summer, adult bears, usually males – twice the size of females – may concentrate on hunting bearded seals, which make a sizeable catch compared to ringed seals and are therefore worth the effort. Some even specialize in hunting bearded seals, perfecting the technique of stalking them in the water. But it takes practice to perfect the skill of sneaking up on an adult seal, especially when that involves creeping through the channels between the sea ice or even swimming under the ice without making a ripple.

As polar bears are their main predators, bearded seals are constantly alert. They even sleep close to the ice edge, with their heads facing downwind and towards the water, sensitive to the slightest warning sound and ready for a quick exit. The pups are born on ice floes near open water and can swim shortly after birth, quickly becoming proficient divers, which is likely to be an adaptation to being hunted by polar bears.

▲ **It's behind you.** The polar bear (left) makes a silent approach – a stalking skill that takes practice. The bearded seal senses danger but fails to see where it's coming from.

▶ **1–6 The lucky and the unlucky.** In one powerful move, the bear has lunged out of the water and onto the ice – not quite fast enough to grab the seal but skilfully enough to dive after it and catch it.

WHEN CHICKS FALL LIKE RAIN

By July all the Arctic's visiting birds are starting to head south again. Soon the ocean, which they have relied on for food, will be freezing over, and the long days of summer will revert to darkness.

The ground-nesting birds are all on small islands or coasts and have chicks that can make the short dash to sea almost as soon as they hatch, even if they can't fly, and some, such as eider chicks, gather together on the sea for protection from marauding gulls.

The geese and the seabirds that nest on cliffs have a greater challenge. Their chicks have no option but make a treacherous leap off their rocky eyrie. They can't fly, and so a lot of them just bounce their way down the cliff face. Most survive but then risk falling into the mouths of Arctic foxes. This is the boom time for the foxes, and they know it. They wait for the falling chicks all day long and catch so many that most end up cached underground as a winter store.

The guillemots have devised a clever technique to minimize their losses. All their chicks are synchronized to fledge over just a week or two. In that period, the sky seems to rain chicks – trying to glide to safety and the sea. Their parents fly down with them, delicately adjusting their descent by pulling on their chicks' tails. Unfortunately, not all the nesting cliffs are up against the ocean. Often the chicks find themselves crash-landing on the

tundra or the beach. Then they have to make a mad scrabble to the sea, running the gauntlet of glaucous gulls and foxes.

The adults stay beside a grounded chick to see off attacks from gulls, but they can do nothing when a hungry fox comes around and may themselves fall prey to it. But with so many chicks jumping at once, the foxes are quickly spoiled for choice, and most of the fledging guillemots make it to the sea. From there they swim slowly south for several weeks before they are able, at last, to fly.

▼ **Eyeing the heavyweights.** A young polar bear looks longingly at a group of huge walruses on a beach on Svalbard. Its only chance would be to stampede the herd and try to grab a baby crushed in the chaos. But it would be a dangerous move, as walruses will defend their young and are capable of killing a bear.

THE LAST-CHANCE WALRUS

By the start of autumn, the vast majority of the summer bird visitors have headed south, and only 18 animal species remain. All of the Arctic predators have to prepare for a long winter with almost nothing to eat.

Life becomes increasingly difficult for polar bears, even those that have managed to catch enough seals to put on weight for the winter and especially for the young inexperienced bears. The sea ice has completely melted, and all the seals are at sea. They have one last desperate hunting attempt at the end of summer.

Walruses like to haul out on ice floes, but by the end of summer, when all the ice has melted away, they are forced to haul out on the beaches, gathering in colonies that may be many hundreds strong. Even a big polar bear can't deal with an adult walrus, with its thick hide and lethal tusks. But walrus pups are easier prey, though somehow a polar bear has to separate a pup from the main herd.

One technique is to run at the hauled-out walruses, hoping to cause a panic as the colony rushes for the water, sometimes crushing a pup or leaving it behind. But adult walruses don't give up their pups easily, and polar bears can be killed in the ensuing battle.

THE AUTUMN GLOOM

Autumn in the Arctic has nothing of the excitement and drama of the spring. There is no sudden return of the sun or transformation of land and sea. Instead the autumn has a brooding, unsettled feel. The wind gets cooler, the storms are more frequent, and eventually the sun slips below the horizon. Even the freezing of the sea ice is less dramatic than the melt. An oily stillness comes across the surface of the ocean until thin, frozen pancakes of ice start to form. Finally these are packed together by the wind, and solid ice begins to form.

Polar bears wait impatiently for the frozen ocean to be strong enough to carry their weight. Over the summer they have shown extraordinary adaptability to the ever-changing conditions. Unlike most large predators, which use the same hunting techniques all year round, polar bears have to use totally different techniques with every passing month – and it gets harder every time. But now, at last, winter is here, and times will get a little easier for the king of the ice.

CHAPTER SIX
HUNGER AT SEA

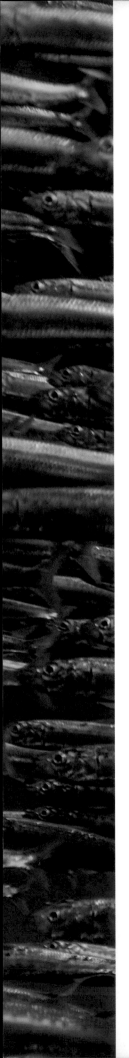

THE OPEN OCEAN COVERS MORE THAN
70 PER CENT OF THE EARTH'S SURFACE,

and for the most part, it is a lifeless desert.
Where life does exist, it is patchy and
unpredictable, shifting with the seasons and
the currents. The predators that do survive
here are among the most spectacular and
specialized, adapted to travel massive distances
using minimal energy in search of unpredictable
sources of food. For their prey, there is nowhere
to hide in this world without walls. Some gather
in giant shoals to gain safety in numbers.
Others are transparent or reflective to try
to disappear into the blue. The deep ocean
beneath is even vaster. It makes up more than
80 per cent of the planet's living space. The
deeper you go, the less food you find – and
the competition between predators and prey
becomes ever more intense.

▶ **Mass movement.** Sardines swim in synchrony, relying on safety in
numbers to counter predators attacking from both above and below.

◀ **(previous page) The interface.** Huge frigatebirds skim the surface picking
off fish chased up by underwater predators. They can't land on the sea.

THE DRIFTERS AND THE SWIMMERS

Far out in the open ocean, the only source of energy is the sunlight trapped by phytoplankton. These minute plants make up half of the planet's living matter and produce more than half of the atmosphere's oxygen. Sunlight doesn't penetrate far below the surface, and so the majority of ocean predators live within the top 30 metres (100 feet). All of them are constantly on the search for areas rich in phytoplankton. The magic combination of sunlight and nutrients enables the phytoplankton to bloom. The sunlight varies with location and season, while most nutrients, which come from the deep ocean, are brought to the surface by rough weather or upwelling currents. This means that plankton blooms are, for the most part, unpredictable and ephemeral – and predators have to be permanently on the move to find them.

There are two approaches to this journeying lifestyle – planktonic and nektonic. The plankton go with the flow, putting themselves at the mercy of the wind and currents. These powerless wanderers range from the smallest zooplankton to the largest jellyfish – and even the 2000kg (4400 pound), 3-metre-long (10-foot) sunfish, the biggest bony fish of all. The nekton, by contrast, swim. Far less abundant than the plankton, they include most of the fishes and squids, the turtles and sea snakes, the penguins and of course the marine mammals, including the great whales.

◄ **The world's biggest mouthful.** Having lunged through a great swarm of shrimp-like krill with a gape big enough to encompass a bus, the blue whale has engulfed an enormous amount of water, expanding its 'throat' pouch to accommodate it. The krill are filtered out through the curtains of hair-like baleen that hang down from its mouth.

WORLD WITHOUT WALLS

It is hard to imagine lion cubs being on many predators' menus, but in the ocean, the young of most of the big predators start life at the bottom of the food chain. Even marlin and tuna begin as larvae just a few millimetres long, and though they're already voracious hunters, of copepods and other smaller plankton, they too have to avoid hungry mouths. The challenge for all oceanic plankton is where to hide when there are no burrows to disappear into or coral heads to duck behind. In this world without walls, the strategy is to be as inconspicuous as possible. And that's why almost all plankton are transparent.

Pteropod snails have adapted to life in the liquid blue by changing the foot into two transparent wings, which makes them look like tiny ocean angels. Larvaceans float in a see-through house of mucus, with an open end so water can be drawn in over mucous filters that capture phytoplankton. If these filters become clogged, the house is thrown away and a new one created. Some of the most efficient planktonic predators are the largely transparent jellyfish,

▲ **(left) Predatory jelly sac.** A 10cm-long (4-inch) transparent ctenophore, or comb jelly, propelled by rows of hair-like cilia that, as they move, scatter the light in rainbows of colour. As it swims, it swallows smaller planktonic animals.

(centre) Snail as sea butterfly.
A pteropod, or sea butterfly, eggs in tow. It's effectively a snail with a transparent, uncoiled shell – in this species, 10mm (0.4 inches) wide, with three spines. Propelled by its foot, developed into swimming wings, it catches its prey with a mucus net.

(right) Sea sapphire. A copepod, carrying two sacs of eggs. It is one of very many shrimp-like copepod species that make up at least 60 per cent of the marine plankton biomass – food for many.

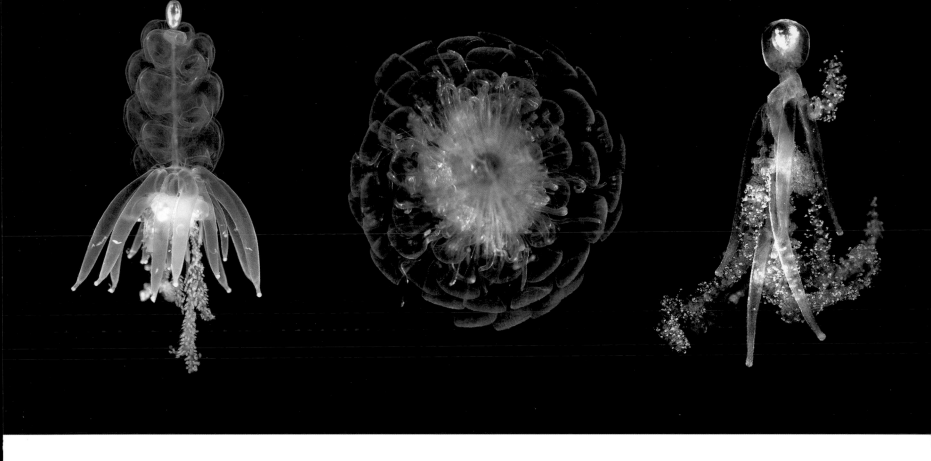

▲ Carnivorous social jellies. Each is a siphonophore – a colony of groups of individual animals organised as a whole predatory unit. Like their relative, the Portuguese man-o'-war jellyfish, they use stinging tentacles to catch prey.

(left) Hula-skirt siphonophore. Its float and swimming bells can be up to 12cm (5 inches) long. At the bottom of the float is a gas-producing pore that controls the buoyancy of the colony.

(centre) Rosy siphonophore. Instead of one umbrella bell, it has multiple 'petals', each a repeating unit specialized for swimming, feeding or reproduction.

(right) *Rhizophysa* siphonophore. A 10cm-high (4-inch) jelly. When fishing, its tentacles may stretch to nearly a metre.

such as the lion's-mane jellyfish, with its pulsating bell carrying masses of powerful stinging cells for stunning and dispatching prey.

Among the most beautiful transparent wanderers are the ctenophores, or comb jellies. In a light of a diver's torch, the rows of fused cilia (hair-like structures) that run along their bodies set off a firework display of colours. Ctenophores have a range of hunting techniques. The small sea gooseberries use two long sets of sticky nets to capture copepods, while the much larger beroe, which looks like a flattened spaceship, hoovers up bigger prey including other comb jellies.

The longest of all the gelatinous planktonic predators are the siphonophores. Some of these are superorganisms – colonies several metres long made up of four different polyps with different roles. *Physalia*, or the Portuguese man-o'-war, has one large polyp adapted into a gas-filled float that catches the wind and propels it along, and hanging from its body (a colony comprising two different polyps) are metres of tentacles (another kind of polyp) covered with stinging cells that create a transparent wall of death.

HIDING WITH LIGHT

Practically all the nekton animals – the ones that propel themselves – are carnivorous. They include the plankton-eating fishes – herring, sardines and anchovies – that gather in shoals hundreds of thousands strong.

The largest fish in the ocean – the 14-metre (46-foot) whale shark – also feeds only on plankton. So do those great whales that filter seawater through massive plates of baleen – the baleen whales. But most of the nekton feed on each other in classic food chains – smaller fishes eaten by bigger ones. These include the ocean's speed merchants – the tuna and the billfishes, the sharks and the largest nekton-feeding whale, the sperm whale.

The muscles these predators need to swim mean they can't play the transparency trick. Instead, many of these fishes use counter-shading. Seen from above, they are dark blue, and from below, silvery-white, like the light playing on the surface. Many fishes also have reflective silver sides that help them disappear in the blue. The mackerel has taken this a step further with stripy markings to break up its body shape and so confuse predators.

Most of the nekton predators have large, efficient eyes, but even in the clearest tropical seas, visibility is limited. Water, however, carries vibrations well, and most fishes have lateral lines along their bodies that pick up the tiniest movements and changes in water pressure. These lines may also play a key role for fishes that gather in massive shoals to protect themselves from predators. Just as wading birds twist in a coordinated flock when attacked by a peregrine, sardines in a giant shoal do much the same, and lateral lines may be crucial for synchronizing their movements.

Sharks and rays have yet another sensory system – organs (ampullae of Lorenzini) that have a range of functions including touch, salinity and magnetism and can detect weak electrical signals created by muscle movements as well as changes in water temperature that indicate where prey species might be found feeding. In a hammerhead shark, these ampullae are at either end of its hammer-shaped head, which it's thought might help it pick up faint magnetic fields on the seabed, allowing it to follow magnetic roads.

◀ **See-through silver shoal.** Indian mackerel, almost translucent, power through the water of the Red Sea, filter-feeding on tiny plankton animals near the surface. Their reflective silver scales and side stripes help break up their body outlines, and they practise safety in numbers, congregating in shoals that can be thousands strong.

LONG HAUL, LOW COST

The ocean's top predators have to be prepared for long journeys to find their food – a satellite-tagged tuna crossed the Atlantic in just 119 days, averaging 65km (40 miles) a day. These tuna, billfishes and sharks may go many days without food, and so have streamlined torpedo-shaped bodies to reduce as much as possible the drag caused by the water and therefore the amount of energy they use. Most have no scales and eyes covered with transparent lids. Their stiff, narrow fins tuck away in grooves along their sides, and many have keels near their tails that direct water flow efficiently across their bodies. Long journeys need powerful engines, and these predators have large, thin, sweptback tails that produce maximum thrust for minimum effort. They are also packed with muscle rich in myoglobin, which stores the additional oxygen they need.

To keep their muscles warm, these fishes use a counter-current system: cold veins run parallel to warm arteries so that the blood is warmed as it returns to the heart. This system works so well for the giant bluefin tuna that it can venture into water as cold as 7°C (13°F above freezing), while still maintaining a core body temperature of 25°C (45°F). This allows it to hunt in food-rich, cold waters, far from the tropics where all the other tuna species have to remain. Swordfish have warming muscles around their eyes which keep them 4°C (7.2°F) warmer than the surrounding water. Warm retinas can process information faster, giving this hunter speedy reactions. It also makes the swordfish's eyes more sensitive in the low-light conditions hundreds of metres down where they like to hunt.

FAST FOOD AND FASTER FISH

The ocean's top predators need more than just stamina. When they finally find their prey, they need to be sprinters, producing a short burst of intense speed to complete a kill. Recording the top speeds of fishes is notoriously difficult, but the best estimates are: yellowfin tuna, 75kph (47mph); wahoo (a torpedo-shaped tuna relative), 77kph (48mph); and the sailfish – the fastest marine predator – 108kph (68mph). Largely solitary, a sailfish can also keep up a constant cruising speed of more than 48kph (30mph). Like the other billfishes – marlin and swordfish – it combines a muscular, streamlined body with a long, tapering bill.

Billfishes, though, will work together to predate a shoal of smaller fishes. They will quickly raise and lower the giant fins on their backs to frighten the

▲ **Snapping up the last sardines.**
Striped marlins work as a group to round up a shoal of Pacific sardines. The energy they expend in their fast-moving attacks is compensated for by the number they can catch in one hunt. As they sprint in, their side stripes flash ultraviolet, almost certainly confusing the terrified little fish and possibly signalling to each other as a way to avoid collisions.

little fishes into a tighter swirling ball – a baitball. Once they have their quarry where they want it, the billfishes coordinate their attacks to avoid damaging each other, taking it in turns to make high-speed runs at the panicking shoal. The phosphorescent stripes on striped marlin flash ultraviolet as they make their breakneck turns – a light show that might help avoid collisions and certainly confuses the prey (many fishes' eyes are highly sensitive to ultraviolet). As they strafe through the baitball, the hunters stun their prey with rapid flicks of their bills and then just suck down the dazed fishes.

◀ **Foraging formation.** A gathering of many pods (social groups) of spinner dolphins, hundreds strong, travel in the late afternoon to an offshore foraging area. They tend to feed on prey such as shrimps, squids and lanternfishes that live deep down but migrate up every night to feed nearer the surface.

▶ **(top) Travel mode.** Spinner dolphins increasing their speed of travel by porpoising – leaps alternating with coasting and swimming – much as penguins do, cutting down on water drag. They often travel in a line, literally sweeping the ocean with ultrasound.

▶ **(bottom) Synchronized swimming.** Spinner dolphins revealing their streamlined shape, built for speed. They keep in contact and coordinate their movements using whistles, echolocation and visual cues.

SOLITARY OR SOCIAL

A lone operator will get the entire prize to itself, when that prize is eventually found. But in the vast ocean, searching together, as tuna do, makes it easier to find the prize. The ultimate social hunters have to be the dolphins, which are armed with an advanced prey-detection weapon – an echolocation system that uses high-resolution sonar. Dolphin groups can be hundreds, even thousands, strong. Leaping out of the water to spot distant prey, they coordinate a hunt by constant communication.

Every evening the spinner dolphins that live around the Hawiian islands leave the shallow inshore waters where they have been resting, safe from sharks, and head out to sea to hunt. Their favourite prey are the shrimps, squids and myctophids, or lanternfishes, that spend their days at depths but migrate into shallow waters each night. Foraging dolphins hunt in pairs within pods of about 20 that may gather together in a group hundreds strong. Using echolocation, the group may 'sweep' the ocean in a line up to a kilometre long. They corral their prey, circling in on it, coordinating their formation. Once the prey is dense enough, the encircling pairs take turns to feed in an orderly way, communicating all the time, so that all the dolphins benefit.

These dolphin feasts attract other predators such as tuna and billfishes. Indeed, tuna shadow-hunt dolphins. Once the shoal is driven close to the surface, seabirds such as gannets and shearwaters will dive in to take their share. When in full action, a dolphin-created baitball feeding frenzy can be one of the natural world's most impressive predation spectacles.

▲ **Biggest takes all.** With one last huge mouthful, a Bryde's whale takes and virtually finishes off a sardine baitball.

◀ **Dive attack.** Shearwaters join the sealions to attack the Pacific sardines from above, while the tuna attack from below, keeping the shoal at the surface.

◀ **(previous page) Feeding frenzy.** Sealions attempt to feed on the swirl of sardines. Surprisingly, the sealions were not very good at making the most of the plenty, and the sardines' silver-shoaling worked as a defence. Only when the skipjack tuna attacked from below and shearwaters from above did the sardines form a tight ball that was easier for sealions to attack.

SAFE IN A CROWD

It is not just the predators that know the value of working together. Many prey species can also see an advantage in numbers. Half the fishes in the sea regularly form schools as juveniles, and a quarter continue this behaviour throughout their lives. Herring form the largest schools of all, coming together in what could be the greatest gathering of a single-species animal. One shoal in the North Atlantic contained more than 3 billion fish.

There are a number of advantages for prey species to gather in this way. In the first place, even the smallest fishes are predators themselves, and their foraging improves when many eyes are working together – as does their ability to spot approaching predators. When the shoal is under attack, coordinated movement can confuse a predator and a silver wall of tightly packed fish can seem impenetrable. And for any individual fish in the shoal, the risk of being eaten is diluted by sheer numbers.

THE OCEANIC AIR FORCE

The endless search for food in the open ocean doesn't just happen below the surface. Above the waves, some of the bird-world's most specialized hunters scour the surface in search of a meal. You can sail for hundreds of miles across the Southern Ocean and see nothing, and then out of nowhere, a wandering albatross will arrive beside your boat. Even in the vastness of the churning seas, these great white birds always impress with their size. With wingspans exceeding 3.5 metres (12 feet), the wandering and royal albatrosses have the longest wings of any bird. But what is truly remarkable is how, for hour after hour, they will follow in your wake without ever seeming to beat their giant wings.

▲ **Running launch.** Black-browed albatrosses try to take off after feeding on krill close to the surface of the sea. With such long wings, this requires a huge effort and a long run to get enough air under the wings to give them lift.

Their wings aren't just long but also narrow – up to 15 times longer than wide – perfectly adapted for exploiting prevailing winds with two energy-saving techniques, dynamic soaring and slope soaring. Dynamic soaring relies on the fact that the surface of an angry sea reduces the speed of the wind above it, creating a gradient that an albatross can ride. First, the bird flies into the wind, gaining height by angling its wings until it reaches stalling speed. Then it turns and, with the wind behind it, descends, using the tailwind to gain speed. Using this technique, it can travel thousands of kilometres without flapping its wings.

Slope soaring is more straightforward. Updraughts from the massive waves that roar across the Southern Ocean give height. Long, thin wings also give high glide ratios – for every metre an albatross drops, it travels forward 22. It can achieve speeds of 127 kilometres per hour (79mph) and can keep up a fast speed for more than 8 hours, hardly ever flapping its wings. To save even more energy, a special tendon locks the outstretched wings in place.

The Southern Ocean is the roughest on the planet, and its constant violent churning brings nutrients to the surface, making it surprisingly rich in places. But these hotspots are unpredictable, and so, more than almost any other seabird predator, albatrosses have to travel enormous distances to find their prey. The results from satellite tracking are astonishing. A grey-headed albatross that left its breeding island off South Georgia in the South Atlantic circumnavigated the Antarctic continent in just 46 days, and a male wandering albatross, nesting on the Crozet Islands, went on a 10,000km (6,200-mile) trip to find food for its incubating partner, returning to the nest in just 14 days. The researchers were also able to monitor the energy the wanderer used on his epic journey and discovered that he had only used twice as much as his partner as she sat patiently on her egg. It seems that albatrosses steal 80–90 per cent of the energy they need from the endless winds of the Southern Ocean.

Albatrosses are members of the petrel family – birds that have tube-like nostrils which almost certainly give them an extra-special (for birds) sense of smell. It's thought that the swarms of krill or phytoplankton at rich upwellings give off an odour that petrels can detect from far away. Albatrosses' long wings keep them from diving far below the surface, but their sense of smell probably helps them find anything floating. But once on the surface, there's another drawback to long wings: taking off is energetically demanding. Overall, though, the energy used by a wanderer searching for prey is only twice that of a resting bird. By comparison, gannets with their spectacular diving technique burn six times as much energy as a bird on the nest.

Albatrosses use the wind, but they're also its prisoners, and on rare calm days, they sit becalmed on the surface. So dependent are they on strong, reliable winds that all but 4 of the 19 species are restricted to the roaring fifties and forties in the southern hemisphere. The only one that lives in the tropics is the waved albatross, which breeds in the Galapagos Islands. This species and the three that live farther north make up for relative lack of wind by relying on closer, more reliable food supplies that require shorter foraging trips.

▶ **Oceanic wanderers.** A wandering albatross rides the air waves, practising a see-sawing dynamic flight pattern, using the updraughts of South Atlantic waves, all the while looking for fish, squid or even jellyfish near the surface of the sea. The search for food hotspots may take it on a journey of thousands of kilometres, and a foraging trip can last more than a month. Albatrosses may also scavenge, whether food caught by other predators or offal from ships.

◀ **Hunters and pirates.** A frigatebird hovers over a likely source of food – the remnants of a shoal of sardines decimated by a gang of sailfish in the Gulf of Mexico. The likelihood is that some sardines will be close enough to the surface or even leap out of it so they can be snatched by the frigatebirds, which can't land on the water, having sacrificed waterproofing in favour of lightness for soaring. But the main food of frigatebirds is squid and flying fish, which flee to the air when chased up by predators below. They also practise piracy, attacking other birds to force them to drop their food.

FLYING PIRATES AND FLYING FISHES

The clear blue waters of the tropics are far less productive than seas at higher latitudes, with their rough weather and upwelling zones. The prey of tropical predators is scarce and spread over a large area. One avian predator, though, has taken up this challenge more than any other and has become the albatross of the tropical seas. The frigatebird is a threatening black silhouette of a bird with extremely long wings. It has the lowest wing-loading ratio – the area of wing compared to total body weight – of any bird, and it has done everything possible to reduce its weight. It has even jettisoned the special oils that all other seabirds need to stay waterproof.

Frigatebirds need to be light for low-energy soaring on the weak thermals that form when the trade winds cause convection off warm tropical seas. They are therefore limited to the trade-wind zones that cover 40 per cent of the globe on either side of the equator. Much like albatrosses, they have an orientation of their shoulder bones that allows their wings to be outstretched for long periods without using much energy. Soaring on thermals, these pirates cover vast distances, with a successful feeding opportunity only every 105km (65 miles) on average.

Frigatebirds seem to track specific patterns in the air currents that correspond to patterns in the sea's warmth and movement. These lead the birds to the fronts in the ocean where phytoplankton bloom and their prey tends to gather. Frigatebirds also have a close association with predators such as tuna, dolphins and dorados. Soaring at a great height and with excellent eyesight, they can spot the little explosions at the surface that reveal where other hunters are at work. But lacking waterproofing oils, unlike all other seabirds, they can't risk landing on the surface. Instead, they steal food from others. They're the fighter pilots of the high seas, and they can outrun other seabirds, forcing them to regurgitate their catches.

Another potential food source is flying fishes. These torpedo-shaped fishes shoot out of the water, propelled by a powerful tail beating up to 70 times a second, and fly above the surface to escape hungry jaws below. Their two elongated pectoral fins function as wings, and a smaller second set near the tail turns them into biplanes. As gravity drags them back to the surface, the long tail dips in to give another boost. They can repeat this trick more than a dozen times, covering hundreds of metres in just a few seconds. An unlucky few, though, are snapped up on the wing by frigatebirds.

DESCENDING INTO THE DARK

Any predator that dares to leave the warm, sunlit surface waters to hunt in the deep ocean immediately faces enormous challenges. Only 20 per cent of the energy fixed in shallow waters will make its way below about 30 metres (100 feet). Even in the clearest tropical waters, sunlight barely penetrates deeper than 150 metres (490 feet), and photosynthesis is no longer possible below this depth. Oxygen becomes increasingly sparse until a point, at 500 metres (1,600 feet), when all the oxygen created at the surface has been totally used up by other animals. It is a frontier most predators are unable to cross. Temperatures plummet quickly, too, and below 1000 metres (3300 feet), the ocean is a uniformly cold 2°C (35.6°F). Most demanding of all, with every 10 metres (33 feet) you descend, the pressure increases by one atmosphere. So at just 500 metres down you have to deal with pressures 50 times greater than at the surface. Even for humans, journeys into the deep ocean are more demanding than space travel. That's why so much of the abyss remains unexplored.

There are, however, a few determined predators that will risk visits to the deep in search of prey. The blue shark is known as the cold-water shark because of its ability to survive down to 7°C (45°F). But though it can dive to at least 1250 metres (4100 feet) in search of squid, it can only remain at this depth for a short time before it has to return to the surface water to warm up again.

Sailfish are the deepest-diving billfishes. They have large, sensitive eyes, behind each of which is a specially warmed muscle to keep everything working at depth. Of the turtles, most stay in shallow waters, but leatherbacks can go to 1300 metres (4265 feet) looking for jellyfish. Unlike the other turtles, leatherbacks have highly flexible shells that don't fracture under pressure.

The record-breaking diving predator, though, has to be the elephant seal. It can go down as deep as 1500 metres (4921 feet) and stay there for up to two hours. Its thick blubber keeps it warm and is full of oxygen-rich blood, and though its lungs are crushed out of action at just 40 metres (130 feet), it can reduce its heart rate to just six beats a minute. In this state of torpor, it can spend more than an hour searching for deep-sea squid.

▶ **Deep-blue diver.** A blue shark hunting in the epipelagic zone – where light still penetrates and where blue gives an element of invisibility. When it is hunting for squid, it is capable of deep dives. But without adaptations for the cold, its dive-time is limited and it has to return to warmer water – also necessary to speed up digestion.

◄ **(top) Amphipod** – a shrimp-like deep-sea crustacean, with enormous transparent eyes. Under cover of night it travels up to the light zone to feed on even smaller planktonic creatures, running the gauntlet of its own predators.

(bottom) Clawed armhook squid – with a see-through body covered with chromatophores (pigment-containing, light-reflecting cells). It can change colour instantly to blend with its background or to signal to other squids. A skilled hunter, it can move incredibly fast, both forward and backward.

► **Viperfish** – with an expandable jaw and fangs that form a trap from which it would appear there is no escape – an adaptation for the depths, where prey seldom passes by. The light at the end of its dorsal spine (not visible here) may help it lure in prey, and the photophores (light-producing organs) along its belly are used for counter shading in the gloom of the twilight zone.

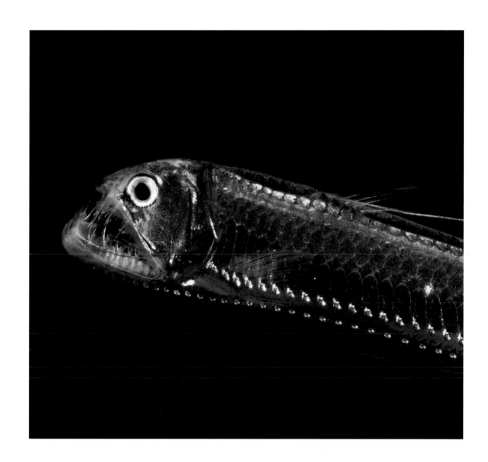

DOWN TO THE TWILIGHT ZONE

In clear tropical waters, a tiny amount of sunlight can penetrate down to 1000 metres (3300 feet). This gloomy world is called the twilight zone. Animals here never meet a hard surface and so don't need a skeleton. This allows many to hide by becoming transparent. The shrimp-like *Cystisom* amphipods, for example, are see-through jewels, with two large transparent eyes for penetrating the gloom. Even complex animals such as squid and octopus have become transparent in the twilight zone. Those few organs that can't be made transparent, such as some eyes, are camouflaged behind a reflective layer of silver.

Predators here need very sensitive sight. Many fishes have tubular eyes, designed to be able to look up to spot silhouettes. Several shrimps and even an octopus have this same design. The deep-sea *Histioteuthis* squids have two different eyes – a large one that looks up and a smaller one that keeps watch below.

Many animals in the twilight zone have disappearing acts, but the real Houdinis of the gloom are the hatchetfishes. Not much larger than a postage stamp, these are paper thin with perfectly reflective silver sides that transform them into mirrors of their surroundings. They are so thin that they hardly make a silhouette from below, and even this is camouflaged. Along their narrow bellies are light-producing organs – photophores – that can adjust their brightness to match the sunlight filtering down from above. On a bright day the photophores glow strongly. On dull days they are less active. This allows the hatchetfishes to disappear at any depth.

A wide range of animals living in the twilight zone use photophores for counter-illumination. But some predators have a way to crack the camouflage – massive eyes with yellow lenses that allow them to distinguish between light from photophores and light from above.

With little detritus drifting down, the twilight zone is thin on food. Rather than waste precious energy, some predators just wait for prey to come to them. Sea spiders, for example, catch copepods by holding out all their extremely long hair-covered legs, paraglider-like, and sieving the little animals out of the water. The long, slender cutlass fish holds itself bolt upright, waiting for prey to come by, which it catches with its long, pointed snout and sharp teeth. Snipe eels are very, very long and very thin, delicate fishes with bird-like beaks, the two sides of which curve apart and are covered in tiny hooked teeth for catching shrimps. As they swim they sieve prey out of the water.

Many of the predators, however, need more than the slim pickings of the twilight zone. Each night millions of tonnes of them and their prey rise into the richer, shallow waters – the largest predatory migration on the planet. Many animals are involved, but the most numerous are fishes – lanternfishes (the most numerous vertebrate in the world). Just 5–15cm (2–6 inches) long, they are all muscular swimmers, and unlike many fishes from the twilight zone, they have swim bladders and can adjust their buoyancy with depth.

The length of the vertical migration depends on the animal. Tiny plankton may journey just 10 metres (33 feet). But some lanternfishes, named for the photophores that cover their bodies, will take more than three hours to travel from way down at 1700 metres (5577 feet) to just 100 metres (330 feet) below the surface. Under the cover of darkness, these diminutive twilight predators can probably avoid the eyes of larger shallow-water competitors. As dawn approaches, they will descend to the relative safety of the twilight zone.

▶ **(top) Snipe eel** – a scaleless fish from the twilight zone. It has more vertebrae in its back (at least 600) than any known animal and can be up to 1.3 metres (51 inches) long. The snipe eel appears to feed on tiny crustaceans and may swim like a snake, its mouth open, snagging them on its backward-pointing teeth.

(bottom) Lanternfish – an 8cm (3-inch) lanternfish, or myctophid, with glowing photophores (paired light-producing cells). The lights may serve to confuse predators by breaking up the fish's outline and may also be used for communication. Lanternfishes migrate up at night from the twilight zone towards the surface to feed, and in their turn provide food for many other predators.

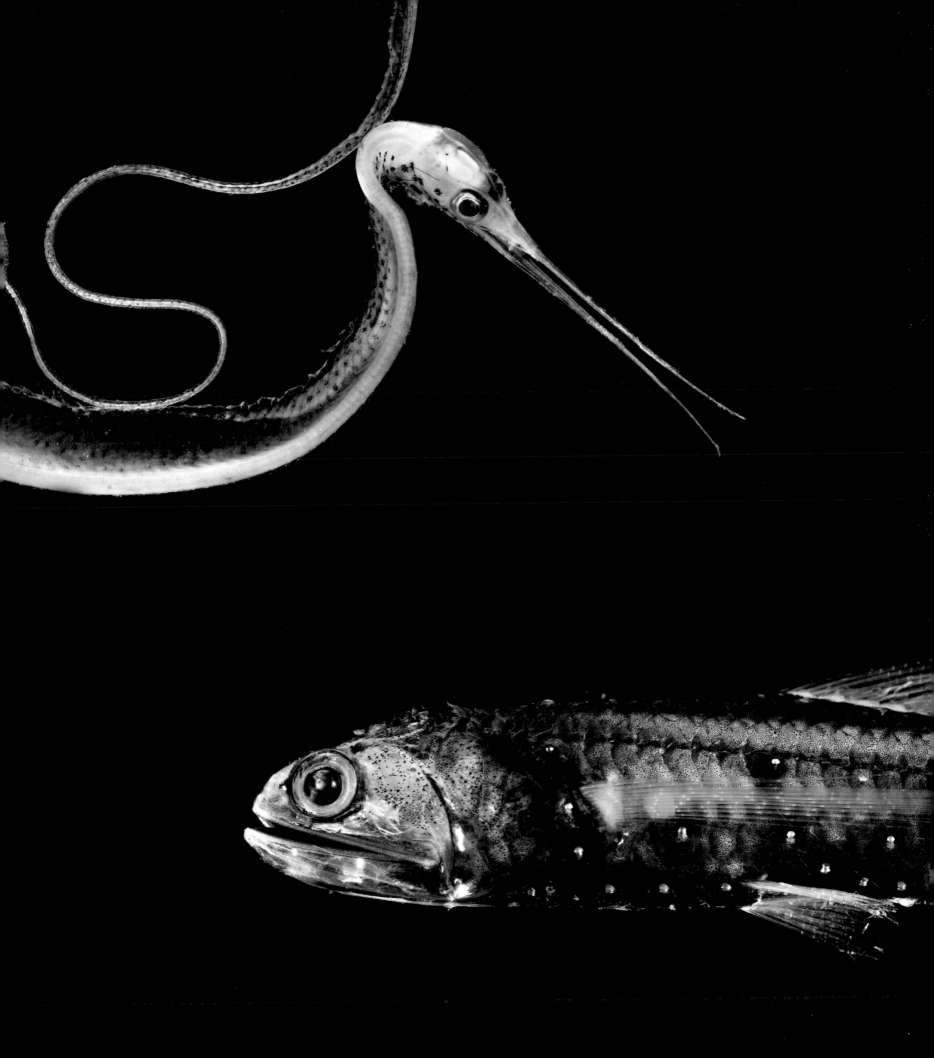

◀ **(top) Big-fin squid** – just 10cm (4 inches) long. Little is known about this deep-sea hunter or what the glandular structures at the ends of some of its tentacles are used for. It has numerous chromatophores (pigment-containing, light-reflecting cells) on its body and tentacles, probably used for communication as well as disguise.

(centre) Glass squid – a tiny, round (1.5cm) see-through predator that can pull its head and arms into its mantle cavity and fold back its fins to form a turgid ball, presumably for protection, aided in some way by a covering of tubercles.

(bottom) Ghostly seadevil – an 8cm (3-inch) female anglerfish that lacks both pigment and the luminescent fishing-rod lure that most anglerfish have, though it has a classically very big mouth. Its spines are presumably for protection.

DEEP INTO THE DARK ZONE

Journey deeper than 1000 metres (3300 feet), and the conditions for predators become yet more demanding. The pressure here is 100 times that at the surface, temperatures hover around 2°C (35.6°F), and absolutely no light from the surface penetrates down. This is the dark zone – a vast space that makes up more than three quarters of all the water on the planet. Only 5 per cent of the energy created in shallow sunlit waters makes it down to the dark zone. For the predators here, distance and pressure changes make it impossible to migrate into richer waters near the surface, and so they are adapted to a world where the pickings are very thin indeed. They are also among the most bizarre animals on the planet.

It's hard to imagine a weirder family than the anglerfishes. Their names alone – like blackdevil and triple-wart seadevil – give you some idea what to expect. Most of them are surprisingly small, just a few centimetres long, since there is simply not enough food to support many larger animals. They are completely black – perfect camouflage in a world without light. Other creatures are dark red, because blue water quickly absorbs any red light from above – dark red in effect is the new black. Because they never meet a hard surface, anglerfishes have weak bones and flabby muscles, and their eyes are tiny because there is little to see in the dark.

But anglerfishes are sensitive to the smallest vibrations. The hairy angler is completely covered in sensitive antennae. This 'listening station in the dark' looks like a large, hairy beach ball. All anglerfishes have highly extendable stomachs and huge mouths lined with terrifying teeth. Prey comes along so rarely that they must be able to tackle any size meal. The gulper eel has taken this wait-and-see technique to the extreme – its body consists almost entirely of mouth and highly extendible jaws. It hangs vertically in the water column waiting to engulf its prey in its enormous umbrella of a mouth. And it can handle prey bigger than itself.

Anglerfishes get their name because they use bioluminescent lures. More than 90 per cent of the animals in the deep ocean create their own light this way. It's done by burning a substrate called luciferin, using a catalytic enzyme called luciferase. On top of its head an anglerfish has a pole with a lure filled with symbiotic bacteria that create the bioluminescence for their hosts. Because light is so rare in the dark zone, an anglerfish's flashing lure attracts attention. Inquisitive prey that come too close are snapped up.

◄ **Small-tooth dragonfish** – showing the light-producing photophore below its eye. It also has a long fishing rod (barbel) hanging from its chin with a blue-flashing lure at the end. Most deep-sea fishes see and use blue light, but this dragonfish can also see in red light and its red photophores can cast an invisible (to most) red spotlight on its prey.

▶ **(top) Siphonophores** – small deep-sea siphonophores from the Atlantic, glowing with bioluminescent light. Like all siphonophores, they are colonies of individual animals, each with a specific function in a complex whole.

(bottom) Helmet jelly – a large jellyfish (with a bell up to 35cm high) living at depths of 7000 metres (23,000 feet). It moves up at night to feed on plankton (such as the copeopod, just visible top right), which it catches using the stinging cells on its tentacles. Being red, it is invisible in the depths, but it can produce rapid waves of blue bioluminescent light up and down its body to startle predators.

The range of fishing rods is extraordinary. Some anglerfishes have poles three or four times longer than themselves. Many also have bioluminescent barbels on their chins made up of branching filaments that resemble bizarre Christmas decorations.

Other predators that use bioluminescence include dragonfishes, a group of small deep-sea fishes with long, thin bodies and armoured heads full of fang-like teeth. The glowing barbels beneath their chins come in all designs. One 15cm-long (6-inch) dragonfish even has a barbel that hangs 2 metres (6.5 feet) beneath its jaw. Quite how prey attracted to this lure are snapped up by such a distant mouth remains a mystery as no one has ever seen it happen.

Dragonfishes have got one more weapon. Behind their eyes, they have a powerful photophore that acts a little like a searchlight. They are also muscular by deep-sea standards and obviously are prepared to seek out their prey. One family – the loose-jawed dragonfishes – has gone a step further. The photophore behind the eyes produces red light instead of the normal blue. Because no red light penetrates to these depths, most of the animals here can't even see red light, which means these dragonfishes can secretly search out their prey.

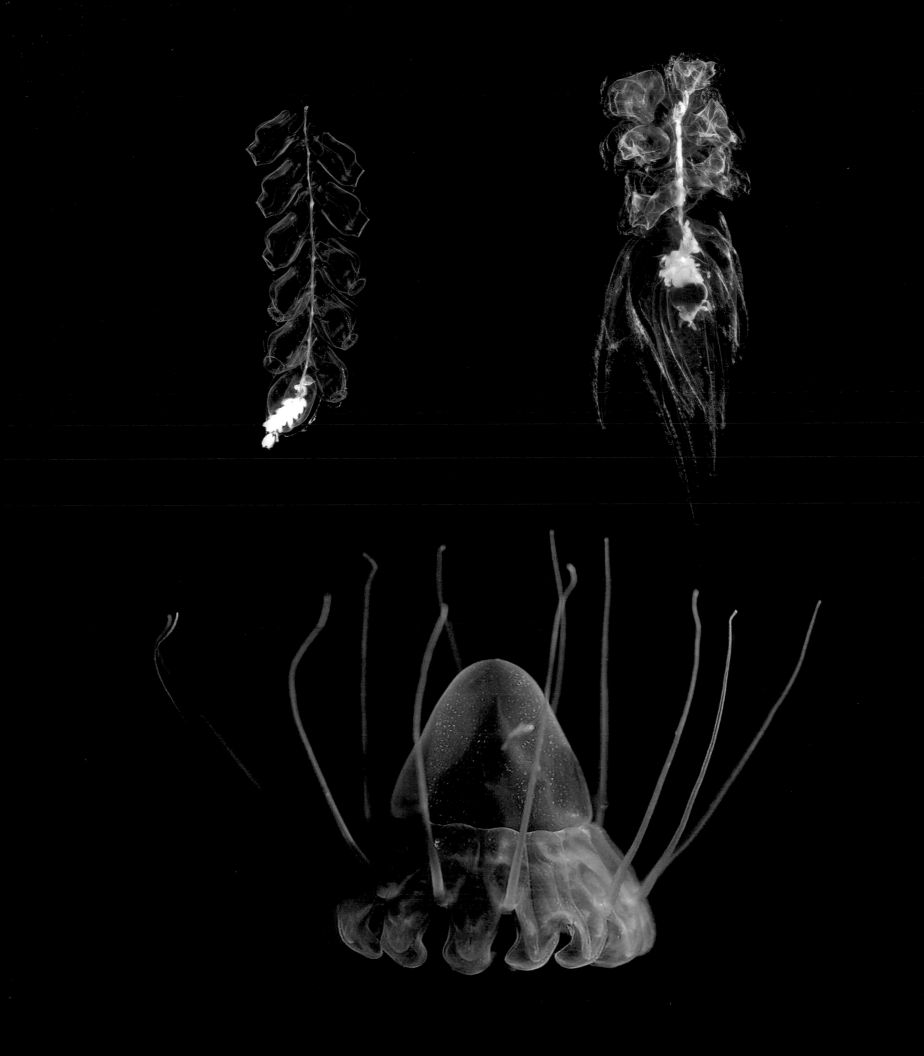

(top) **Slimline-traveller.** A 25-metre (82-foot) blue whale in search of krill. Swarms of krill may be great distances apart, and so a blue whale is constantly on the move. It is even presumed to give birth in transit rather than on breeding grounds. Streamlined for energy-efficient, long-distance travel, its huge tail fluke is a powerful means of propulsion.

(bottom) **Size efficiency.** Having dived down and then lunged up at speed through the remains of a krill swarm, the blue whale's accordion-like 'throat' has expanded to take in a volume of krill and water greater than its own body mass. The krill are filtered out as the water is expelled through curtains of hair-like baleen (keratin) that hang down from its mouth. Though a blue whale has to travel great distances to find large-enough aggregations of krill and lunge-feeding uses a lot of energy, the feeding technique is highly efficient, with a 90 per cent gain in energy, enabling the blue whale to support its huge body size.

THE LARGEST PREDATORS OF ALL

The largest animal ever known to have lived on the Earth is the blue whale, weighing up to 175 tonnes. Like all the great whales it is a long-distance traveller, constantly searching the ocean for the enormous quantities of food it needs. That search takes many of the great whales to higher latitudes for the polar summers, where long days of almost continual sunshine make the Antarctic and Arctic seas among the richest. Each summer, the silence of the sheltered bays all along the Antarctic Peninsula is broken by the blows of hundreds of humpbacks. But when winter returns and the ocean freezes over, the whales head back to warmer latitudes to breed. These journeys towards the equator can be 8000 kilometres long (5000 miles), and when the humpbacks reach their destination, there is little to eat in the lifeless tropical waters.

These large long-distance ocean predators need near-perfect streamlining to reduce the clawing drag of water. The blue whale is perhaps the most streamlined, with a relatively slim body, elegantly drawn out to nearly 30 metres (100 feet). Its massive tail fluke delivers thrust with 90 per cent efficiency – far more than even the best ships' propellers – and in a short burst, it can reach more than 50kph (30mph). Humpback whales are about 18 metres (60 feet) long, with a far less streamlined, almost dumpy shape. But they do have the largest pectoral fins of any cetacean – up to 5 metres (16 feet) long – which enable humpbacks to do their famous leaps. The leading edges of the fins are studded with knobbles called tubercles, which redirect the flow of water over the tops of the fins, increasing lift.

The favourite prey of blue whales, humpbacks and many other great whales are shrimp – krill. If the largest predators on the planet are to rely on such small prey, the krill need to be harvested in enormous numbers. A blue whale feeds on nothing but krill and is thought to consume more than 40 million in a day. Luckily for the baleen whales that feed in this way, krill remain one of the most numerous animals in the ocean. Estimates vary, but there are thought to be more than 500 million tonnes of them in the Southern Ocean alone. But even these swarms are hard to find in the vast expanse of the open ocean. Nobody is quite sure how blue whales find the swarms of krill they need. It does not seem to involve echolocation, and though males seem to communicate when they are travelling, they are usually silent when they feed. One theory for this silence is that whales may

be listening for their prey. Certainly crustaceans are famously noisy, and a giant swarm might make a real racket.

Even when the whales have found a swarm of krill, not every swarm will do. Blue whales and all the other baleen whales have evolved a unique form of hunting called lunge-feeding. They have loosely articulated jaws and highly expandable throat pleats, which they can open into a cavernous mouth. Then they can engulf a whole swarm in a single, spectacular lunge. Only from the

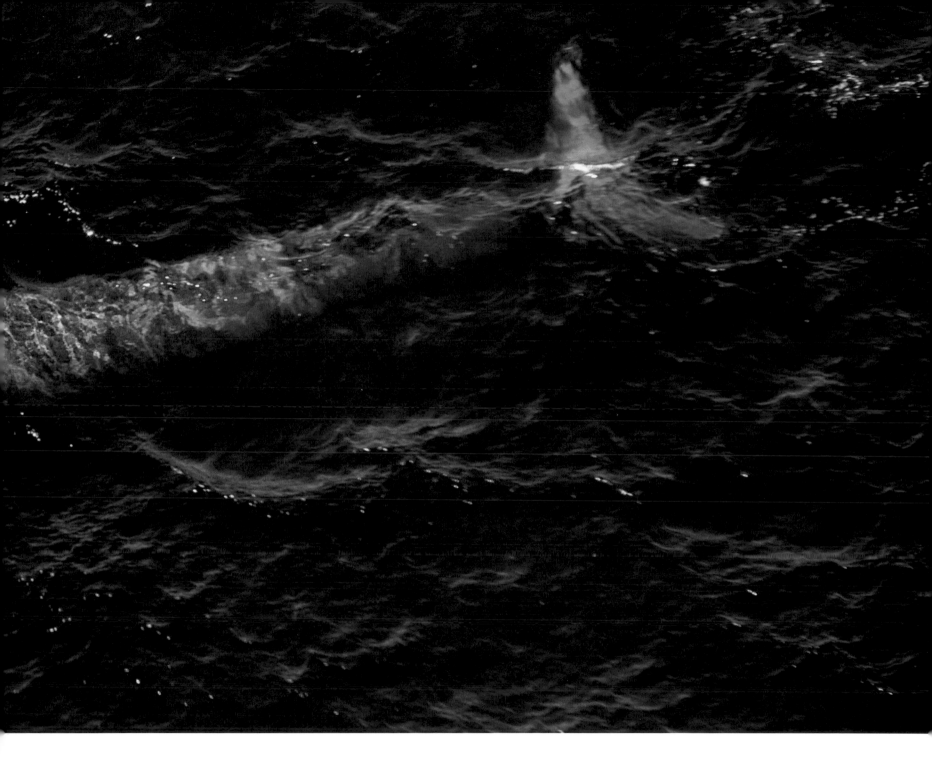

▲ **The biggest single breath.** A blue whale takes a breath after a ten-minute lunge-feeding dive. Its exhalation blow is more than 6 metres (20 feet) high. On a high-energy dive, a blue whale rapidly uses up its oxygen, and so the longest it can spend at depth is about 15 minutes.

air can you really appreciate how this transforms the whole shape of a blue whale. Its once long and sinuous body now has a giant balloon on its front, full of hundreds of gallons of krill soup. As soon as it can, the whale draws in the expanding pleats and sieves out the krill by forcing the water back into the ocean through its baleens. This process of lunge feeding is energetically demanding for a whale, which is why it will swim right past less dense swarms and only stop for the richest meals. The largest predator on the planet is a picky eater.

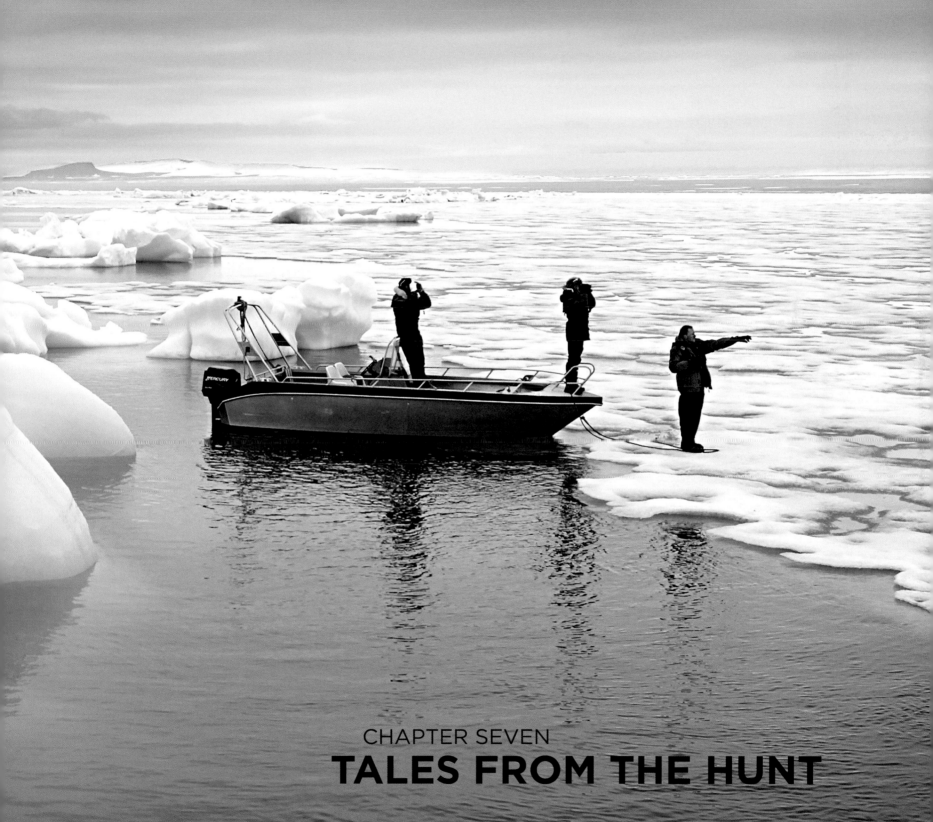

CHAPTER SEVEN
TALES FROM THE HUNT

'IT WAS AN AMAZING 45 MINUTES – WE WERE WRUNG OUT WITH ADRENALIN AND CONCENTRATION,' says producer Ellen Husain recalling the attack of killer whales on a humpback mother and calf. 'Then I realized… Wow! No one has ever witnessed a battle of this magnitude, let alone filmed it.'

'In just ten minutes, we'd got the Holy Grail,' says series producer Huw Cordey, on capturing an entire cheetah hunt from the air – 'the luckiest thing that has ever happened to me in 20 years of filming.'

'It was the most amazing experience of my career,' says producer Hugh Pearson, describing being circled by a jumbo-jet-sized blue whale.

Again and again, returning crews reported new behaviour, new insights, new experiences. So you could say that the production team was incredibly lucky. But then luck is about being in the right place with the right team at the right time.

▶ **The biggest shoot of all.** Cameraman David Reichert films off California as the world's largest animal, the blue whale, swims past with a mouthful of krill.

◀ **(previous page) Ice-bear watch.** The polar bear team off Svalbard searching for hunting bears swimming amid the melting ice.

'It's all in the title,' says *The Hunt*'s executive producer Alastair Fothergill (producer of some of the most memorable of wildlife television series, including *Blue Planet* and *Frozen Planet*). 'It's not about the kill but the hunt – the strategies and the struggles, the failures as much as the successes.'

It's also about taking calculated risks to get the strongest stories and a determination to film a story in a way that will create an emotional connection – a sense of being there with the animal, in the action – and editing it to show the hunting strategy unfold.

THE CINEFLEX ANGLE

One of the most valuable tools for revealing the strategies has been the Cineflex – the gyro-stabilized, high-definition camera system that allows smooth, steady sequences. But its application when filming wildlife requires both ingenuity and expertise. Jamie McPherson was the camera operator tasked with devising different rigs to operate the huge, delicate Cineflex on boats, jeeps and even an elephant, to show the animals from new perspectives and with different styles and to reveal behaviour impossible to film with a camera on a tripod.

The aim was always to make the shots cinematically beautiful – 'to the Fothergill standard', says Jamie. 'I didn't want viewers to notice how anything

◀ **The tiger angle.** Jamie McPherson aboard Gotham (see page 245) heads off into the forest in India's Bandhavgarh National Park looking for the tigress and her four cubs. He could be gone for several hours, following the tigers as they move deep into the forest.

was filmed,' but to feel 'immersed in the sequence' as if they were in the pack of dogs running with them, gliding with the polar bears or racing with the spinner dolphins.

In the case of the African wild dog hunt, it meant using a specially strengthened vehicle, rigging it with scaffolding and a jib arm off the side, plus a dampening device, to carry the Hollywood-style camera that would allow smooth shots while driving at up to 65kph (40mph) alongside the dogs on rough ground. It also meant locating an area (in this case, a remote region of Zambia) relatively free of termite mounds and aardvark burrows (hitting a hole could destroy both camera and vehicle) and having a rally driver on the team.

'The dogs' pace was frenetic,' says Jamie. Though they'd start off at a trot, they would speed up as soon as a wildebeest was targeted. 'We'd be with them for ten to twenty minutes as they ran at speed, filming in real time.'

The shots are 'so incredible that it looks as if the dogs had been trained to run alongside the car', says series producer Huw Cordey.

▲ **Running with the pack.** Two younger members of the pack of 13 African wild dogs pick up speed chasing a wildebeest. Despite the rough ground, Jamie was able to film alongside, using the stabilized Cineflex hung from a cushioned rig. For the dogs, the greatest danger of speed was the possibility of broken legs.

▶ **Nowhere to hide.** The wildebeest, the dogs and the filming jeep, seen from the helicopter in the late afternoon. From 9am to 4pm, the dogs could be reliably found at one of the waterholes. As soon as they'd eaten, they'd go to a waterhole to drink and rest during the heat of the day, before going hunting again. The night crew would then keep track of them until dawn, when filming could resume.

SKY VIEWS

At the same time, the team was filming aerials, using a Cineflex and an ultra-long lens mounted on a helicopter. Again they struck lucky, filming a whole hunt to complement the ground shots and to show how the wild dogs break down a herd of wildebeest and single out an animal. What allowed the deconstruction of the dogs' hunting strategy in the edited sequence, says Huw, 'was that combination of being able to move between the top shot, where you really understand the distances, and the ground, where you can see the stamina and speed'.

The helicopter offered different challenges, including just keeping the subjects in frame while being buffeted by the wind. When you film from a normal camera on a tripod, 'you become part of it', says Jamie. 'You can see everything with the left eye and scan with the right.' But when operating a Cineflex, 'you're looking into a monitor hood, and there's a slight delay – a big deal when pulling focus. Essentially you're controlling it with a motor down a wire… panning and tilting not with your body but a joystick. At the same time, you're also directing the pilot – about altitude or position, not left and right – or talking to the producer, and all through a headset. So you really have to concentrate… The producer's looking at a monitor – and with a tendency to go "wow!" just as you are pulling focus on a running animal.'

When filming Arctic wolves chasing zigzagging Arctic hares on the tundra, it was impossible for cameraman Mark Smith on a quad-bike to keep up with the chases and take more than chunks of the action at ground level. But Jamie could take over from above and show the hunt in one go.

Of course, what makes the final edit look extra-special is that much of the series' footage was shot in *ultra*-high definition – that means cinema quality and, in some cases, IMAX-cinema quality. And when you can use the Cineflex to show in one frame a whole wolf pack chasing a muskox herd, that *is* a truly cinematic experience.

STALKING A TIGER

For the tiger hunt, the requirements were completely different: slow, smooth tracking at eye-level. But in India's Bandhavgarh National Park, off-road driving is forbidden, as is getting out of the vehicle to set up a tripod. You can strap a tripod in the middle of a jeep, but then it's not possible to film

▲ **Grounded.** The Zambia crew waits for the dogs to wake up. Flying the helicopter was hugely expensive, and so every day that they couldn't film was a worry. From left to right: cameraman Jamie McPherson, pilot Frank Molteno, series producer Huw Cordey, field assistant Robin Dimbleby and assistant producer Mandi Stark.

down low or to make a 360-degree turn of the camera. So once again, Jamie needed to build a special rig and a jib arm for the Cineflex so it could track smoothly at various levels alongside a stalking tiger, turn at any angle and swing over the driver to the opposite side. For off-road, low-down forest scenes, producer Jonnie Hughes came up with the idea of building a special aluminium elephant rig with a winch for the camera. Carrying it and Jamie would be Gotham, a 65-year-old tiger-tracking elephant,

who had no problem being fitted with the rig and was completely at ease with tigers.

Says producer Jonnie Hughes about filming from the jeep in thick forest: 'Quite often, at our level, we couldn't see the tiger, but Jamie could, standing up and looking through the Cineflex, and could tell us to keep driving, directing us... No way could we have followed her otherwise... and when you are moving parallel to her, the shots are unbelievably beautiful.'

It was in the seventh week that the team had their reward – the first-ever start-to-finish film of a tiger hunt. 'We picked her up walking, knowing that the chance of seeing her hunt was tiny,' says Jamie. 'Ahead of her were two chital stags, walking in the opposite direction. When she disappeared behind a bush... I took a gamble, focused on one of the browsing stags... Then suddenly, she burst out of the bushes and knocked it straight over, like a rugby player. There was no struggle, not like a lion on a wildebeest – no fight. In just 30 seconds, she smothered it and slid it behind a tree.'

'I didn't see it,' says Jonnie, 'as I was bent double pushing the jib up in the air. Jamie was looking at the screen. All I heard was "damn". She'd gone behind a tree... At that point we put the camera on the elephant and recorded the rest of it. To film a kill of that quality in a forest is remarkable... and to have the two camera ideas working together was amazing.'

◀ **Jumbo transport.** Gotham waits patiently as Jamie makes sure that the camera is working. The Cineflex had been transferred from the jeep in preparation for following the tiger family into the forest. The bespoke aluminium rig allowed it to be winched up or down to tiger-level and the camera angle to be controlled with a joystick. When being filmed, the tigers took no notice of Gotham, which allowed intimate shots.

▶ **The final sprint.** With a chital deer in her sight, the 18-month-old female cub bursts out of the bushes but fails to catch it (a frame from the video shot with the Cineflex mounted on Gotham). The most adventurous of three cubs, she was learning fast from her mother how to hunt.

AQUATIC-STALKING GAMBLES

On a boat, even a big one, there is a lot of movement, and so it is difficult to make the slow cinematic tracking moves needed for travelling dolphins or a polar bear stalking a seal in the water. Certainly, when you spot a bear, you can't jump out onto the ice with a tripod – and it's not just a matter of interfering with the bear's behaviour. There are quite big safety problems. So once again, a way had to be found to use the gyro-stabilized Cineflex – the only way to achieve

the low-angle shots needed to show the skill of a polar bear in a maze of shifting ice. The solution was to fix the camera on a crane that could be balanced on a small metal skiff, which could be lowered quickly into the water. Then you have to find a hunting bear in a vast environment. 'In drift ice, its head is like a small white duck,' says Jamie. 'If you are lucky to find a bear aquatic stalking, you have to hope it's a hungry one' – and keep your distance. The team followed several polar bears through the broken ice, each one targeting

▼ **The melting landscape.** Having recorded from the boat a polar bear's seal hunt, Jamie and producer Jonnie Hughes, with guide Håvard Festø, stand on the ice to film the fast-melting backdrop.

▲ **Drifting in search of a bear.** On a mirror-still sea, with the engine off, the filming boat drifts as the team looks for seals or swimming polar bears. They found both among the ice close to the glacier ahead. Jamie operates the camera and field assistant Andy Bedwell works the jib, which can be lowered for a water-level view. Standing is location manager and polar bear expert Jason Roberts.

◀ **Thin Lizzie on the move.** Desperate to catch a seal, a female that the team named Thin Lizzie moves in the direction of possible prey. As she got nearer, she would take to the water and, with only her head showing, would swim almost motionlessly amid the ice.

seal after seal. The bears could swim without leaving a wake and could hide motionless behind a piece of ice while weighing up the exact position of a seal, 'then swim under the water, surface 20 feet away and leap on the ice in one move'. But every time the seal got away. After two weeks, the team finally filmed a successful hunt. The water was mirror still, the light was perfect and the boat was able to drift alongside the bear. 'We thought the seal had got away, but the bear obviously followed it under the water and then came up with it in her mouth – an amazing thing to see.'

The special crane arm went on to enable the filming of a lot of the special ocean sequences, whether on a boat following a mother humpback and her calf racing to get away from killer whales – topside shots with real dynamism – or alongside albatrosses at sea or skimming frigatebirds, providing smooth pans of them soaring and gliding.

GETTING DOWN VERY LOW

Technology is changing so fast that there is always a better camera to give sharper, closer pictures. But it is still the filming perspective that determines what technology is needed and how it should be used. To film a rainforest superpredator required a fresh view – both to convey the enormity of the spectacle and the detail of exactly what was happening.

The solution was to use a brand-new 4k mini-camera, designed to film engineering processes, adapt it to take special quality lenses, including a new microscopic one, place it on a pan-and-tilt head, build a 3-metre-long crane jib and invent a device to control the camera focus remotely.

The location was upriver in the Ecuadorian Amazon, and the chosen spot was a kilometre from the camp – the furthest that it was possible to haul all the gear to every day. The predator's trails were obvious. On day one they found the bivouac – the giant organism's camp – and on day two they set up to film the marching army, from the front as it advanced into virgin territory. Cameraman Alastair MacEwen operated the pan-and-tilt head, producer Jonnie Hughes operated the jib, and Luke Barnett operated the focus. They started by filming in front of the swarm. 'The head was more than 7 metres across,' says Jonnie, 'a mass of black army ants, scaring everything out of the undergrowth. You realize how much is under the leaf-litter as everything gets the hell out… even vipers and jaguars are scared of army ants.'

Standing still, their rubber boots topped with bands of Sellotape to stop ants running up any further, the team escaped attack, and the hoard funnelled around them. And using the jib, they were able to fly the camera over the swarm and swing back within seconds to focus on anything happening in the radius of the swarm, such as a cricket being dismembered (see pages 80–1), or to look back along the river delta of ant columns – impossible to do with a tripod, which would have involved running alongside the swarm or kneeling in it.

Another specialist camera allowed them to get close to single ants and to slow down the movement to see what was happening and to give an idea about how fast they move. 'First we shot 60 frames per second, but it wasn't slow enough to stop the blur,' says Jonnie, 'then 90 frames a second, then 120 frames. What was amazing was that you then could see that the ants are making choices where to step, carrying stuff heavier than themselves while trying to avoid their team-mates running in the opposite direction.'

▶ **Shooting soldiers.** Cameraman Alastair MacEwen goes macro with the army ants, while producer Jonnie Hughes and researcher Ilaira Mallalieu illuminate the tiny subjects with LED lights. For wide shots of the marching swarm, the team used a special crane and cabling that allowed operation of the camera remotely. But for close-ups of a soldier, the only way was for Alastair to kneel in the path of the ants and get bitten – a lot.

DIVING INTO THE BLUE

It took a year to develop and trial a custom-built polecam. This involved fitting the latest, smallest ultra-high-definition camera into an underwater housing that could be strapped to a boat and operated on deck via cables linking it to a monitor. Cameraman Doug Anderson would use this to adjust the focus and zoom. It was an investment of time that paid off. Without it the team would never have achieved the near-perfect open-water, sometimes split-level, tracking shots of hundreds of spinner dolphins travelling at speed beside the boat. Of course, expertise was needed to locate an open-ocean site where it might be possible to film such scenes. In this case it was 65km

▼ **Hunt party.** A glimpse of some 2000 or more spinner dolphins hunting for lantern fishes in the Pacific Ocean off Costa Rica. It's a frame from video shot from a polecam hung over the boat. Wearing snorkel gear, cameraman Doug Anderson and producer Hugh Pearson then filmed the dolphins while hanging from the boat as it tracked alongside the fast-moving community of spinners, becoming immersed in a cacophony of sound.

(40 miles) off Costa Rica. The luck came with the clarity of water and a rare moment of mirror-calm sea. There was also the matter of being close enough to the dolphins to film them. This involved producer Hugh Pearson in snorkelling gear hanging on ropes off the side of the boat and being dragged along while singing to them.

The dolphins were suitably intrigued, and the result was a sequence that shows, says Hugh, 'what it's like to be in the middle of a battalion of singing, buzzing, suckling, mating, acrobatic cetaceans'. Being 'immersed in this dolphin world, a world you just don't see from the surface' was 'like being at a crazy party'.

▲ **Nocturnal encounter.** A frame from a night-shoot showing a brief encounter with a spotted hyena – a potential rival for any kill the leopard makes. The night-shift operation involved Matt Aeberhard filming her with an infrared camera from one vehicle and assistant producer Mandi Stark operating the huge lighting panels from another – not an easy or always successful exercise.

◄ **Start of the night shift.** The 12-year-old female star descends from her daytime resting tree to start night-hunting. She also hunted in the day, which gave the team one of their most spectacular sequences.

LIGHTING AND LEOPARDS

The night-time leopard shoot in South Luangwa, Zambia, should have been straightforward. It was the dry season and very, very hot, but filming at night would be cooler. And, of course, leopards usually hunt at night. The decision was to use an infrared camera rather than thermal imaging, which would give more dramatic and artistic lighting. But that required a second vehicle to light the scene, no headlights and excellent communication between the cameraperson and the lighting operator. Says assistant producer Mandi Stark, in charge of operating the huge infrared LED rig: 'We were out 13 hours, but it felt like 2 it was so hugely exciting... I've never seen so many eyes shining out of the dark... Hearing the sounds of what's going on all through the night was really magical.' That was week one. By week four it had become just 'hugely challenging'.

The ground was rock hard and rutted with animal tracks, which in the dark meant a hazardous, jolting ride and punctures every other night. They'd brought special voice-operated headsets, so cameraman Matt Aeberhard could film at the same time as telling Mandi when and where to operate the lights (12 LED panels, in 3 sections of 4). But the headsets didn't work. The lighting rig was so huge that she couldn't see anything while behind it, and the small infrared camera that was supposed to help guide her and the driver was useless when the rig was turned on. In the end they were forced to communicate via the drivers, whose first language wasn't English – 'all this while trying not to disturb either leopard or prey'.

What saved the shoot was a 12-year-old leopard with an accessible territory and who seemed unfazed by being followed – that and the fact that there was a second crew keeping track of her in the day. It turned out that she preferred to hunt in the day, perhaps because in the extreme heat the lions and hyenas that might steal her kills weren't so active. Of the hunts that Matt and Mandi did manage to film, there was one moonlit stalk in particular that was unsuccessful for the leopard but unforgettable for them.

'I was lighting her to my right, alongside the vehicle,' says Mandi. 'She stayed still for ages, watching intently. The impalas were 50 metres away, feeding. This time I could see her in the moonlight, her ears twitching, crawling slowly and absolutely silently towards them. The anticipation was intense... Then one of them finally saw or smelt her, alarm-called, and that was it. She just got up and walked off.'

▲ **Daylight leopard watch.** While Mandi and Huw search for their star, Jamie sits in the back, watching through the viewfinder, and Thomas the driver waits patiently. The Cineflex was at a leopard's head height to give a hunter's-eye view.

▲ **Daylight leopard hunt.** The leopard moves along the gully without being seen, watching for any prey grazing near the edge that she could ambush. It was a successful strategy, and the team managed to film more than one kill.

The daylight team – Huw and Jamie using the Cineflex on a jib arm tracking less than a foot above the ground – also filmed her. In the day, her strategy was to use a gully to stalk pukus and impalas grazing at the edge. 'It was wonderful to watch as she crept round corners,' says Huw. 'She had to be very careful about territorial male pukus, who would alarm-call... and the yellow baboons... If a baboon crossed the gulley, she would turn and run... and hide round a corner until the disturbance died down.' Finally she got past the vigilant pukus and baboons and brought down an impala. 'In less than ten seconds, she had dragged a pregnant impala nearly double her weight into the gully as if it was a blanket,' says Huw. The baboons saw her, 'went crazy and started running towards the gully. The next thing we saw was the staggering impala coming back out of the gully' – which goes to show that all eyes and ears are against the lone hunter.

DECISIONS, DECISIONS, MONEY, MONEY

'One of the important jobs of an executive producer,' says Alastair Fothergill, 'is risk management.' Filming wildlife is always a gamble, especially when it comes to ocean shoots, where there's a huge risk of sinking a budget and ending up with a big hole in a programme. In any big series, says Alastair, a third of the projected sequences are reasonable risks, and a third can be groundbreaking risks. 'But there's no point taking a massive risk unless there's

▲ **Crab diversion.** Pelagic red crabs off the coast of Mexico being filmed by Doug Anderson. It was a diversion on yet another day without any sign of a sardine shoal. These crabs are eaten by many animals, including tuna, and so it was worth hanging around with them in case big predators turned up. They didn't.

a massive return.' So, for example, 'if you hear reports of killer whales chasing humpback whale calves in crystal-clear water, which could lead to one of the most memorable sequences in the series, you risk spending money on it.'

With so many shoots going on at once, the decision-making was relentless. In the open ocean, says Hugh Pearson, it is famine and then the odd feast. Perhaps the most risky of all his shoots was the blue whale feeding sequence, which had never been filmed before. With blue whale numbers just 3 per cent of what they were pre-whaling, finding one in the open-ocean desert and finding one actually feeding on a krill swarm (itself unpredictable) in water clear enough to film in was a massive gamble. The first year's attempt, off the coast of California, resulted in four weeks of nothing, just a few shots of a blue whale swimming through krill – financially a disaster. But the decision was taken to follow through with the gamble and try again the following year. It turned out to be the best year on record for diving with blue whales, and the plankton blooms weren't dense, giving exceptional water visibility – vital for underwater filming. As so often happens, it was only towards the end of the shoot that they got the sequence: 10 magical minutes of material out of 8 weeks of searching.

A WHALE OF A GAMBLE

It was towards the end of the shoot that they spotted birds on the horizon, signalling krill. Racing over, they found a ball of krill the size of a tennis court being attacked by sardines. Blue whales usually avoid fish, which they don't eat, but Hugh and cameraman David Reichert went into in the water anyway. Then Hugh saw it – a huge shape passing under them and within 2 metres (6 feet) of David. It was a 25-metre (82-foot) blue. 'Then we had the most amazing experience,' says Hugh, 'as it came round four times, in a huge arc, taking bites out of the krill – exhilarating and scary in equal measure.

'If it hit you with one of its fins – the tail fin is more than 7 metres (25 feet) across – you could die, and more worrying, it could actually swallow you… when a blue whale opens its mouth, it could swallow a bus. So the golden rule is don't get in the krill…

'To be in the water within a metre of a blue whale and look it in the eye is something I will never forget, something that few have experienced. We were very, very fortunate. The visibility was good, the sun was out, the camera was working, and we'd got something so unique.'

▶ **(next page) Jackpot.** A blue whale about to feed on a huge mass of krill, being filmed by David Reichert, keeping out of the way of its tail and fins – 'awe-inspiring to share the water with such a huge animal'. To film such a fast-moving giant, Dave had to be dropped off in advance of it on its predicted path.

AERIAL DRAMA

When it comes to money, helicopters are also scary. The locations where they are needed are often great distances from their airports, which means paying for extra days' hire and fuel. In the case of the African wild dogs, aerial filming was essential to show their hunting strategy. But to get to Zambia's Liuwa Plain National Park, the helicopter had to come from South Africa. One week before it was due to arrive, the dogs went missing.

The alpha female had a radio collar, but normally it was only possible to pick up the signal within a 1km (0.6-mile) radius – even less in woodland, which is where the team lost them one night when a storm struck. After a week of searching, they still couldn't be found. Producer Huw Cordey therefore took the desperate decision to pay for a light aircraft to come from Lusaka to quarter the park to try to find the dogs. It failed. Then the helicopter arrived. Costs were now escalating. This was serious.

What finally took everyone's mind off the situation was the appearance of a cheetah near to the camp. Rather than keep the helicopter on the ground, Huw decided to try to film a cheetah hunt – an aerial sequence no one had ever achieved before. Within minutes of being airborne, they had filmed the cheetah chasing an oribi gazelle – 'every twist and turn, in real time, showing the speed and agility'.

The next day, they went up again, and once more had unbelievable luck. 'We'd just got to the area,' says Huw. 'We couldn't spot the cheetah, but within 30 seconds saw a wildebeest running and thought, well, she must be chasing it. Jamie was on to it and said, "I'll just turn on the camera and hope she comes into frame," and she did. We got the whole action of the cheetah running after four wildebeest – two mothers and their calves. One mother virtually reversed into the cheetah, which pulled up but then set off again after the calf. She didn't get it, but we did. In *ten* minutes we'd filmed a Holy Grail – a cheetah hunt from the air.'

For *The Hunt*, successful filming often meant following an unsuccessful hunt. Indeed, more than half the sequences filmed ended in failure. But in the case of African wild dogs, the success rate was 80 per cent, and when the spotter plane finally located the pack again, a long, successful hunt was filmed. And with a whole sequence to cut together, says Jamie, 'we could give an impression of the pace, how dynamic it is, how dangerous it is for the dogs and how scary it is for the wildebeest.'

▶ **Exit dogs, enter cheetah.** Racing into the shot, the cheetah sets her sights on a wildebeest calf. She was being filmed from the helicopter, while the team on the ground frantically searched for the subjects they were meant to be filming – wild dogs. They decided to try filming the cheetah because she was being studied in the same area and was radio-collared. So they knew roughly where she was and could follow her by following her potential prey from the air.

WHEN NOTHING HAPPENS

For Hugh Pearson, there is 'something magical and elemental' about being out in the ocean – 'the real wilderness'. 'I can spend a day not seeing anything and still be happy, because there's always the excitement of the unexpected.' But for most film-makers, when the animals just aren't there or the weather isn't as predicted, excitement wears off. For assistant producer Sophie Lanfear and her team, who went to film polar bears hunting walruses, unexpected sea ice in August (the first for ten years) blocked their passage north, forcing them to stay on the east of Svalbard. Here there were lots of walruses but no bears – rather, in three weeks of 24-hour watching, just one immature male was seen. 'The worst was the boredom,' says Sophie, 'and trying to keep everyone convinced that it was worth sticking at it.'

▼ **The long wait.** Lions on a rest day lie up near a waterhole in Etosha, Namibia. Camerawoman Sophie Darlington and assistant producer Mandi Stark had five weeks of lion rest days before they managed to film a hunt – in the very last hour, on the very last day at last light. In an environment with nowhere to hide, it turned out that the best cover for ambushing prey is provided by storms and darkness.

The last weeks of the dry season were chosen for the lion shoot in Etosha, as this is when all the animals are forced to come to the last waterholes to drink, and when the lions would be more active in the day. Mandi Stark and camerawoman Sophie Darlington managed to locate a likely pride, which was indeed hanging around the waterholes. But it was soon apparent, says Mandi, that the lions' only daylight activity consisted of 'getting up at last light, stretching and then lying down again'. Gradually the team realized that the pride hunted only under the cover of a storm – when whirling rain and sand meant Sophie couldn't risk using the camera. After five weeks with no filmable lion activity, the shoot was a stressful, expensive disaster. But on the very last day, at the last hour, in the last of the light, a miracle did happen.

'When I finally located the lions,' says Mandi, 'the sky was already black. You could feel a dramatic storm front moving in – and you could tell they were going to hunt by the way one of the females started walking. The adrenalin was pumping, and I was shaking to the point I could hardly hold my bins steady... I drove like hell to where there was a signal and I could radio in Sophie.'

By the time Sophie arrived, the females were in position, crouching on either side of a small herd of zebra. 'I literally swung the camera out on the camera port and started filming.'

'Their senses were battered by the storm,' says Mandi. 'The zebra just didn't see the lions... the stallion walked straight past them. The lead lion knew exactly which zebra she wanted, and that zebra knew something was coming.' Suddenly she went for it – 'starting the longest sprint I've ever seen a lion make', says Sophie, but away from the camera and for a kilometre into the near darkness. 'By some miracle I managed to keep it mainly in focus... The most amazing moment was when this light suddenly bathed the lions as they killed the zebra, the most blood-red sunset light I've ever, ever seen... When we got back to camp and loaded it, I realized, oh my God, we've got it, we did it.' So the team finally got their hunt, a very different kind of hunt, and the pride got their meal.

But for Mark Smith, trying to film marine otters with young pups, it really was 'utter disaster'. In the entire two months in Chile, he didn't see a single pup. And though he got just enough topside shots of adults and adolescents, these were only the briefest of glimpses in the sea. 'There's little known about their social set-up or behaviour,' he says. 'When hunting, they are rarely out of the sea and seldom at the surface,' and – unlike the case with most other shoots – 'there is no scientist who can tell you what might happen when.'

CALLING ON THE EXPERTS

For much of the behaviour being filmed, help from scientists made all the difference. In most cases, it was the information in scientific papers that helped them decide what, where and when. Then there were the radio collars fitted by scientists to their study animals, including the Arctic wolves and the African wild dogs, and these helped them keep track of some of the stars.

In the case of the sparrowhawk story, it was the invaluable knowledge of Norwegian naturalist Jostein Hellevik that gave them the shots of male sparrowhawks hunting. He had a feeding station for birds in the forest, which he had been monitoring for years. It also appears to have become a training camp for juvenile sparrowhawks, where they practise on jays. In the ten days of filming, the team didn't see any kills, but they saw jays fronting off hawks. 'Jays are so smart,' says cameraman John Aitchison. 'They run rings round them… They learn from observing previous attacks… and have a brilliant reaction to an incoming hawk, using its momentum against itself. The jay will fly at a log or tree trunk, hit it with its feet and push itself sideways. The sparrowhawk can't react in time and completely overshoots. That would happen repeatedly through the day.'

Replaying every move through the Phantom high-definition slow-motion camera, he also marvelled at the aerial skill of the 'dashing young hawks'. To cancel out the lift from its wings and dive down on a jay below him, says John, 'a sparrowhawk will turn upside down, keeping its head the right way up – a crazy thing to do in flight.'

Says assistant producer Adrian Seymour: 'Though the attack would be over in seconds, the hawks were so regular and our host's knowledge of them so great that we could predict where they would go. So we could plan our shots and even arrange the background. It was like working on a film set with tame birds.'

The Hunt also made very many discoveries just through spending longer periods of time with their subjects, often in places few scientists could manage or afford to get to. For many team members, these discoveries were the highlights of trips and of the series itself.

▶ **Training flight.** A young sparrowhawk attacks a young jay in a practice run at the Norwegian-forest feeding site. Though the male hawk's flight path was accurate – talons ready to grab and stab – the jay was more than capable of outmanoeuvring him.

SPOTTING STRATEGIES

Often it's the cameramen and camerawomen – usually skilled naturalists – who observe subtle behaviour that no one else has, a product of long hours watching and the need to predict behaviour. Barrie Britton spent seven weeks filming breeding birds on Svalbard in the Arctic. One sequence needed was the rain of guillemot chicks jumping from their nests to the sea below. 'Basically the chicks have to take off and clear the whole scree zone at the base of the cliff to hit water. You'd think they would be well developed before they jump, but we saw a wide range of sizes and ages jumping. There seemed no correlation between size and whether they would make it. Sometimes a ball of fluff would go way out over the sea before it hit water, while developed chicks would drop like stones onto the scree. We began to realize that the gulls were hammering the birds on the cliffs, picking off vulnerable chicks. If the chicks waited until they were big enough to have the best chance of reaching the sea, they risked being caught by the gulls, hence the pressure to jump early.'

To make the sequence work, they had to get the moment a chick jumped. But despite hours and hours of watching to try to spot the cue that a chick would jump, they just couldn't work out which nest to put the camera on. Then Barrie had the idea of leaving the camera running on a section of cliff. Reviewing the footage, the only clue was that a chick would walk a couple of inches to the edge of the ledge and face out for 30 seconds (normally the chicks faced in for safety). With its little flippers on the edge, it would then just launch itself off, seeming to catch its parent by surprise.

'As soon as we saw one facing out, we would focus the camera on it. We got better at spotting the moment and managed to get four good shots of a jump.' The glaucous gulls may have been watching for the same clues from their observation perches. But so many chicks would jump in the six-hour evening session that the gulls were soon satiated.

Even more fascinating was the discovery of the Arctic fox strategy for catching little auks. These tiny puffin-like birds were nesting in holes in the boulder scree below the cliffs. Barrie and assistant producer Sophie Lanfear

◄ **Lookout for leaping chicks.** From a precarious spot up the cliff, Barrie Britton trains his camera on the ledges above, trying to spot a guillemot chick making the move that indicates it is about to jump. Watching with him is assistant producer Sophie Lanfear.

◀ **Foxwatch.** Barrie's hide sits among the scree at the bottom of the cliff. From here he would spend the day filming the thousands of little auks nesting among the boulders and the foxes hiding there to try to catch the birds. The area was a natural amphitheatre, magnifying the constant stream of noise from the little auks and the cackle of guillemots above. When the auks took off, there would be a great 'swoosh', and Barrie could feel the energy wave created by their little wings.

▶ **Aukwatch.** A fox watches intently where the little auks are nesting and then hides among the boulders. When a fly-past of a gyrfalcon or gull causes a mass exodus of adults, it will leap up to try to catch an auk, or it will wait patiently until the birds fly back and then try to grab one as it lands.

would see a fox enter the scree and disappear among the rocks, but then nothing seemed to happen. Then one day they realized what was going on. The foxes were hiding among the rocks, sometimes for more than an hour, waiting for little auks flying back to the colony. As soon as one came close enough to the fox's hiding place, it would dash out and try to catch it. 'A dog fox would leap vertically up and try to grab one like a frisbee,' says Barrie. But, says Sophie, the vixen wouldn't risk hurting herself. 'She would wait for an auk to land and, when it wasn't looking, would charge out and grab it.'

SPIDER WAYS

One of the fascinating stories was that of Darwin's bark spider, described for science only in 2009, Darwin's 200th anniversary. This spider sprays silk that, in strength and elasticity, outperforms all other known materials, even other spider silk. It can be 'thrown' huge distances across a river to form the bridging line for a giant orb web. But the precise spraying technique was only revealed when it was filmed in slow motion with a high-quality macro lens.

The behaviour proved very difficult to film. First the team had to find the diminutive spider along a river in Madagascar. Then they had to work out which twig or leaf it would throw its line from. The spider also had to be on eye-level with the camera. In the end, they decided to present it with the ideal starting point – a carefully placed bit of vegetation.

'She hangs from a twig or leaf and points her abdomen,' says Huw Cordey, 'and the silk comes out like smoke. You've never seen anything like it.' The spider sprays it in a fan rather than a line so that the wind will catch it, and what they discovered reviewing the film is that, every second or so, she squeezes the spray together to crimp it. As the silk goes out, the wind spins it, and it starts to form a string. If it fails to catch on the other side, she just reels in the line and starts again. Once it catches in a tree, she reels it in to tighten it, walks across and adds another layer. Then she walks into the middle to make a crosspiece before starting the web. 'Astonishing.'

Just as astonishing was Huw's accidental discovery at the end of the shoot of a spider new to science and with a fascinating behaviour unknown among

▲ **Spider gear.** Thirty-nine bags and the team – cameraman Alastair MacEwen, series producer Huw Cordey, driver Joel and assistant cameraman Ryan Atkinson – in the car park of Madagascar's international airport. 'It was one of the heaviest kit shoots I've ever been on,' says Huw, 'for some of the smallest subjects.'

▶ **Spider shoot.** Local helpers, alongside Alastair and German spider-scientist Rainer Dolch, watch as Ryan sets up the cable dolly – a gyro-stabilized rig – to film the web of a Darwin's bark spider strung across the river.

The pretend spider. The first-ever picture of the web of the first-ever decoy spider recorded in Africa, discovered by Huw in Andasibe-Mantadia National Park, Madagascar. The spider obviously couldn't count – it had made its decoy replica with just four legs.

◄ **The river crossing.** The radio-controlled gyro-stabilized dolly is hung over the river, ready to film Darwin's bark spider constructing her huge web.

African species. As he walked with spider-expert Rainer Dolch along a forest path, Dr Dolch remarked how spiders were everywhere, pointing to one in the centre of a web. Stopping to look at it and noticing something odd, Huw gently poked it, whereupon a much smaller spider jumped out from behind the large one and ran off. 'I realized straight away this was a decoy spider,' he says, 'since we were just about to go to Peru to film one.' As a defence against predators, the small decoy spider uses debris to construct a much larger copy of itself, complete with legs, to sit in the centre of the web. A still picture (above) of the web is the only record so far of the first decoy spider to be discovered in Africa, because when Huw went back to photograph it properly, it was gone.

WHEN PREDATORS JOIN FORCES

The marine shoots often provided firsts – both scientific discoveries and filming firsts, mainly because so little is known about the behaviour of marine animals. Also observations within this vast, hostile (to humans) environment are so difficult to make – and so expensive for scientists to undertake. Indeed, scientists have used BBC footage from big series for their research. But success for *The Hunt* often came only after weeks of anxiety.

For the open-ocean programme, Hugh Pearson and the team set out to film how sardines cope with predators, hoping to get attacks by pack-hunting striped marlin – one of the high-speed open-ocean billfish – but also a range of predators with different tactics. They didn't get marlin, but they did get their baitball.

To remain far out at sea required a boat large enough for a team, including cameramen David Reichert and Doug Anderson, to live on for the three weeks that Hugh calculated would be needed to find the action. The first week went by with little sign of life and a dead-calm sea, and so did the second week. Finally, three days from the end, a radio report came in of possible shoals of sardines two days to the north. The first signs of activity were gulls on the horizon. David and Hugh approached in a Zodiac inflatable and then slipped into the water. At that moment Hugh realized the fins showing in the surface mêlée didn't belong to dolphins but to hundreds of copper sharks. They were frenziedly feeding on a fast-decreasing baitball, also being devoured from beneath by skipjack tuna and from above by sealions. The sharks, in aggressive feeding mode, repeatedly bumped Hugh and David, forcing them to film away from the heat of the action. In just ten minutes, it was all over. As Doug says, it became 'a classic open-ocean shoot – 20 days of absolutely nothing and then 3 days when it went ballistic'. The real jackpot came the next day.

The shoal they found was the size of a large room. Sealions were unsuccessfully trying to feed on it. 'They seemed bamboozled by the thousands of fish.' So it seemed that the shoaling strategy was working. But the dynamic changed with the arrival of shearwaters, dive-bombing from above. Then tuna came, attacking from below, bringing the fish closer to the surface and tighter, and making it easier for the others. Now the sardines' defence started to work against them, as

▶ **Attack from above.** David Reichert films the combined non-stop attack of sealions and shearwaters on the shoal of sardines. The fish were trying to escape to the deep while the predators were pinning them at the surface so they could pick them off.

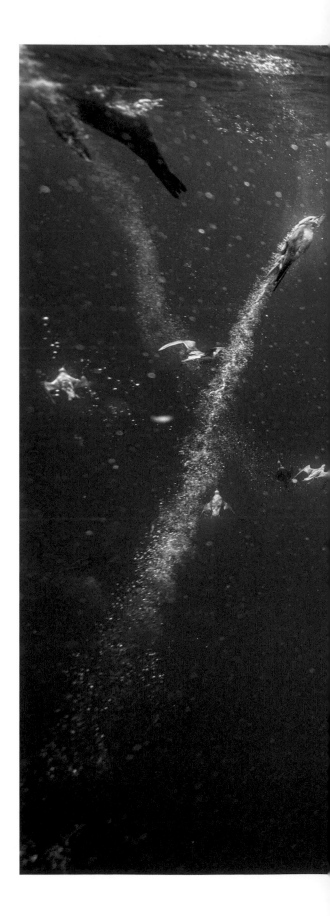

they panicked into an ever-tighter ball. 'You could see them getting tired,' says Hugh. Finally it was as if their brains had scrambled, says Doug. 'They stopped churning,' and when the Bryde's whale powered in, 'it just engulfed them all.'

OTTER FINDINGS

For the coast programme, Hugh was determined to film the marine otter – the world's smallest marine mammal and a hunter truly on the edge. It's found only on the wave-battered Pacific coast of South America (and not to be confused with the much larger sea otter of North America). Here the water is so cold that the little otter has to burn up huge amounts of energy to keep warm and so is forced to live a frenetic life trying to get enough food. It can also be extremely shy – even if you spot one at the surface, as soon as you focus on it, it dives again, which gave cameraman Mark Smith many a problem. And comparatively little is known about its behaviour. Certainly, the chances of diving with one seemed slim. But Hugh finally found a location where otters had been sighted and it was possible to enter the water relatively safely. Here he and cameraman Doug Anderson camped, waiting for a storm-free period when the waves weren't too dangerous and the water was clear enough for filming.

The biggest surprise when spotting his first otter, says Doug, was 'how small it was – half the size of a European otter'. To avoid spooking the shy animals they camouflaged their wetsuits and the camera and used rebreathers to avoid a trail of bubbles from their exhalations. A week of perseverance paid off, and they got the first-ever underwater footage of hunting marine otters.

◀ **Attack from below.** Doug Anderson films a Bryde's whale as it takes a giant gulp out of the remains of the sardine ball, now trapped against the surface.

▶ **Not an otter in sight.** Mark Smith searches for the tiny bobbing head of a marine otter in the surf along the wave-lashed coast of Chile. As he discovered, these little otters have long coastal territories, and when they move up the coast, you can't just pick up your tripod and run after them – one reason they've never featured on TV until now.

Skinny dipping

The smallest marine mammal, the most elusive and the one no one knows much about – not a promising start. And indeed, the team failed to discover anything about its family life or even when it breeds. But they did get something special: the first-ever underwater footage of marine otters, which revealed for the first time how they hunt and one reason why they are so small.

Finding enough food is a real race against time for marine otters. 'They dive down much deeper than anyone realized – 7 to 8 metres – down to underwater boulder fields packed with prey,' says producer Hugh Pearson. 'They disappear into holes, ferret around underground and pop out with a crab or fish.' Underwater cameraman Doug Anderson describes them as 'like little potholers, lithe and skinny. Any bigger and they'd get stuck. The difficulties they have – that they get cold and have to shiver to get warm again – all of it made sense with that one bit of photography.'

▲ **Worthwhile work.** A marine otter surfaces after a deep dive for a crab – a rich find worth enduring crashing waves and extreme cold for.

▶ **1–4 Expert dive.** As waves crash overhead, a little otter dives straight down to the boulder fields, squeezes between the rocks and pops out carrying a small crab.

1

2

3

4

WHEN THERE'S NO NEED TO HIDE

On Ellesmere Island in the high Arctic lives a famous population of white wolves. They aren't hunted and so are unafraid of people. In theory, this gave the team a chance to follow the wolves as they hunted. But when they arrived, explains Mark Smith, the tundra was like a 'muddy Lincolnshire field... We'd set off with the wolves and then spend an hour digging out our quad-bikes and lose them.' Only when the ground finally hardened could they follow them hunting adult hares. But it was so bumpy that Mark had to carry the camera in a backpack to protect it, and a lot of the time he had to stand up on the

Hare stop. Mark Smith stops to film Arctic hares gathered on the tundra. He and producer Jonnie Hughes had spent the day on quad-bikes following the wolf pack patrolling their territory but, having run low on fuel, were returning to camp. The view shows just a tiny part of the pack's enormous Ellesmere Island territory, the still-frozen sea forming one boundary. It has just enough hares and muskoxen to keep the wolves in food.

quad-bike. 'Then you'd find yourself in the middle of the most horrendous tundra hummocks – 2-foot-high bumps as far as the eye could see. The only decision was whether to go incredibly fast or incredibly slow. But either way it would be incredibly painful.' And by the time they had caught up with a wolf chasing a hare running at 48kph (30mph), the kill had happened or the hare had escaped.

But he did manage to get three crucial shots, thanks to the female wolf, who after three weeks seemed to take pity on him, and one evening came and sat nearby and then started to chase a hare, giving him the crucial ground-level

shots. And, of course, overhead was the helicopter and Jamie filming all the wolves' manoeuvres with the Cineflex.

What surprised the team was the selection of prey. When leverets and then adult hares were plentiful, muskoxen were ignored. Indeed, some muskoxen rested almost on top of the wolf den. But there were muskoxen bones all around, and so they were obviously on the menu. And indeed, the team finally filmed a hunt from the air. For Jamie, 'it was the most amazing single battle I've seen.' The team had assumed that the pack would go for calves, but on this hunt 'they singled out a bull' (see page 189). 'They went straight for it,' says Jamie. 'With four wolves attacking, the end was inevitable. But the battle lasted nearly an hour. The male backed into a river. It was horrendous to watch, but in terms of drama, it was all there.'

▲ **Human interest.** Leverets cluster around Mark Smith, sniffing and nibbling. They had no fear of humans – only of wolves – and were always curious, checking out the tents and equipment.

▶ **Camp inspector.** A female wolf trots through camp on her daily inspection. The wolves were incredibly inquisitive and observant, and it was easy to imagine how the original wolf/human relationship could have developed.

THE BATTLE OF THE WHALES

Perhaps the most extraordinary story filmed, in terms of new observations and sheer drama and scale, has to be the killer whales hunting humpbacks off the coast of Western Australia – an 'intense and heart-rending experience' for the team, says producer Ellen Husain. It was also probably the biggest financial gamble of all, with a huge chunk of the budget going on an unpredictable event that even the whale biologists hadn't witnessed.

The first hint of the story was a report in a newspaper. Talking to people on the ground resulted in five reliable witness reports – enough to suggest that the killer whales showed up off Ningaloo Reef for a couple of weeks in July, when the humpbacks were migrating through from Antarctica to their winter breeding grounds farther north. The migrants include pregnant females, travelling to give birth but also mothers who have given birth along the way. Rather than travelling in the deeper, offshore water of the main humpback highway, these mothers and their calves hug the coast and reef edge to avoid the killer whales.

The only chance of filming a hunt was to locate a group of killer whales and then keep track of them. And given the speed and distance they travel, the only chance of doing that was to join forces with killer whale researchers John Totterdell and his Western Australian Orca Research group and US scientist Bob Pitman and his NOAA Southwest Fisheries Science Centre team, who were keen to tag one and observe the pod's behaviour. Darting a whale with a satellite tag was itself a challenge, as the killer whales not only arrived weeks earlier than expected but then disappeared. But they did return, in July, just when hundreds of humpbacks were passing through. And with a tag finally on one of them, Ellen summoned the UK film crew.

The day the crew flew out was also the day the killer whales took off again down the coast, and by the following day, they were out of range of the boat. 'The scientists said: "That's it – they are probably on their way back to the Antarctic," and all I could think was, oh God, I've spent all this money.' Then it went from bad to worse. The crew arrived, the weather deteriorated, and everyone was marooned on land for a week. But just as they were giving up

◀ **Witnesses.** The crew tracks killer whales off the reef. Ellen Husain levels the jib and the gyro-stabilized camera to shoot at sea level, in liaison with, to her left, cameraman Blair Monk, who operates the camera controls from a monitor, while series producer Huw Cordey stands monitoring the behaviour ahead. Driving the boat is David Bond.

hope, the killers reappeared down the coast, heading north again. The boat caught up with them two days later, and the real drama began.

'The pings from the satellite tag indicated they were close in by the reef, where humpback mothers and calves were trying to hide,' says Ellen. 'So we knew something was up. When we got there, we could see a mother and calf surrounded by killer whales. It was high action. Luke [Barnett] was filming from the top of the boat, Doug [Anderson] was operating the underwater camera on a pole, and I was on the monitor. The boat was listing all over the place, and the mother was trying to use it as cover. There was thrashing, heaving and turmoil.' But despite her efforts, the mother lost her calf to the killers. In her desperation, she lifted the boat halfway out of the water. 'In the moment,' says Ellen, 'you are just concentrating on getting the shots. Only when you stop do you realize, oh my God, that mother has just lost her calf.'

They filmed a second hunt the following day. Again the killer whales got the calf (see page 46). 'We carried on following them when they moved off,' says Ellen, 'and Doug went into the water with them.' Just then, 'this whale came like a freight train – trumpeting, steaming out of nowhere – and barrelled into the killers. We couldn't help thinking it was probably the mother.'

By now the helicopter was lined up for aerials, together with Cineflex operator Blair Monk, flown over from Fiji. Using it involved a lot of logistics and expense, including finding ways to refuel along the coast every couple of hours. But the first aerial shoot was the most productive of all. 'It was a mirror-calm day, stunning,' says Ellen. 'We knew that mothers and calves were coming up the reef, and the boat had picked up the killer whales approaching.'

It wasn't until 4pm that the hunt started. 'From the air, you could see their behaviour had changed,' says Ellen. 'They were now on a mission – powering along… There were several mothers and calves around the reef, and we tried to guess which ones they would go for' and therefore which to focus the cameras on. But then the killers split up. 'Two started to harass a lone adult for no apparent reason and followed it for ten minutes like outriders, almost steering it away.'

At that point the team realized another four had surrounded a mother and calf. 'They started repeated attacks from all angles, the mother fending them off, lashing out with her long pectoral fins and huge tail. You could tell how careful they were, dodging and blocking… and from the air, you see how small the killer whales were next to her, how formidable a humpback is. Suddenly two male humpback escorts came in. No one knows exactly what the role of

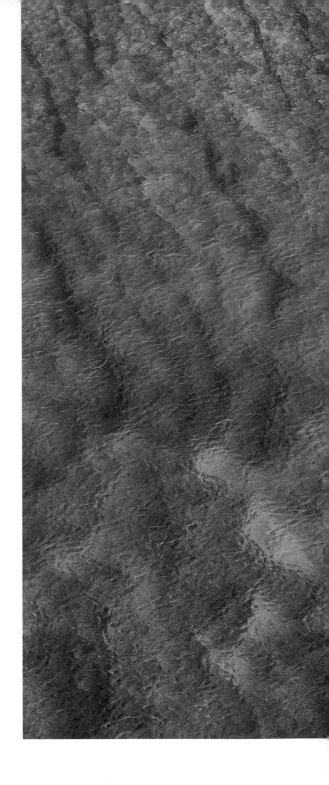

▲ **Safe passage.** Hugging Ningaloo Reef in shallow water, a mother humpback and her calf, accompanied by a large male escort, are trying to avoid killer whales. The main migration is happening in deeper water farther out. Clear water made it possible to film the action from above and, once killers were spotted, to predict where an attack might happen.

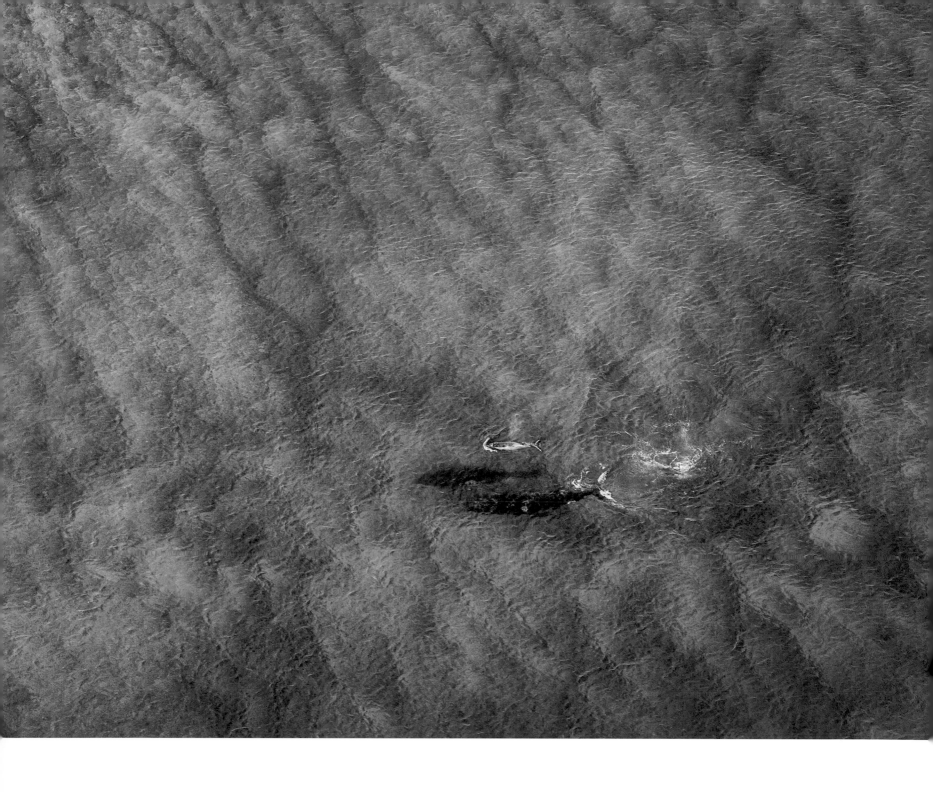

escorts is... But what we filmed was these two males helping to protect the mother and calf. She got the calf up on her back out of the water and then on her head, keeping the attackers at bay. The males would come around and head off the killer whales and swipe them with their tails... always trying to keep the baby in the middle... We were rooting for the baby, and at one moment it got knocked off the mother's back, and she lost it in the boiling water. We thought it was game-over – but then somehow she managed to get

it back up. That went on for ten minutes. Finally, they got away… It was an amazing 45 minutes – something no one had ever witnessed before.'

They saw multiple attacks over two weeks, with a more than 50-per-cent success rate for the killer whales. That, combined with previous reports, has led the scientists to conclude that humpback calves are a predictable and readily taken prey for killer whales. Indeed, they may have even been so over millennia, to the extent that it is possible that these whales need to be big to defend themselves against predators. Doug Anderson says it was 'like watching evolution at work, with the killer whales as the driver'. He believes that the pods they saw may represent one of the small pockets of killer whales that retained their cultural behaviour from pre-whaling times, when humpbacks were plentiful, and that we are seeing the beginning of their population recovery – 'a good way to rethink the traumatic, harrowing and bloody end of a hunt'.

OUT OF THE FRYING PAN INTO THE FRIDGE

Extreme temperatures and extremely long hours seemed to be the norm for most of *The Hunt* shoots. One of the hottest locations proved to be Zambia. Huw Cordey had the daylight shift on the leopard shoot. 'It was hot, really hot, seldom less than 50 degrees… We'd get up at 4am to be in the field for sunrise, because leopards do nothing after 9am, and then we'd go back to camp and just lie on our beds and sweat, before going out again in the afternoon. I've camped in Death Valley, and this was equal to that.' The vehicle had an open back, and so when there was no shade, 'we'd bake… The metal on the rig got so hot that you couldn't touch it – it was like sitting in the sun with a radiator around you.'

Mandi Stark had the cooler night shift filming leopards, but she had to try to sleep in the day, lying in a pool of sweat in the 43-degree shade of her tent. For the few hours the generator was on, there was a fan, but it was the equivalent of 'turning on a hairdryer'. She then went straight to the opposite African extreme – the Ethiopian Highlands – 'so beautiful, but so cold' – to film Ethiopian wolves. 'By morning there would be ice in the tent. I'd sleep wearing six layers: two pairs of thermals, running tights, trousers, waterproofs and a sleeping bag with a hot-water bottle…'

But that didn't prevent it being a wonderful experience. 'The landscape, the animals, the behaviour – it all came together,' says camerawoman Sophie Darlington. 'One morning it was so, so cold – the first really hard, hard frost – that my fingers could hardly work. But the light was coming up… The wolf was curled up on a hay nest made by its prey, with a dusting of frost on its back… Suddenly we heard a howl, and then the wolf pricked up its ears and started to howl. It was backlit… Magical. My favourite shoot ever.'

Cold was a matter of endurance, except when it affected the equipment. Then it was total frustration. Filming Arctic foxes hunting in snow proved to be one of the coldest, trickiest shoots for cameraman Mark Smith and

All-weather fox. One of the film stars, seemingly unaffected by the -40°C Canadian weather. The cameras were not so hardy. One gave up while filming this Arctic fox, and the eyepiece of the second one froze. Cameraman Mark Smith was the coldest he had ever been.

assistant Oliver Scholey. Not only was it hard just to find foxes (it was a fox-population-crash year in the Canadian location), there was also no snow in November, for the first time in living memory. After three weeks, it snowed a little, but then the temperature plummeted to a wind-chill low of −43°C. 'We wanted blowing snow for the drama,' says Mark. 'So we'd go out in it, but the eyepiece of one camera froze up, so I was constantly scraping ice off it, and the other camera just gave up… I was the coldest I had ever been. But the foxes didn't care at all.'

There was just enough snow to use a Ski-Doo to pull the sled loaded with the gear. 'We'd sit on the gear, and the guide would pull us for 30 miles… After an hour and a half of being bounced up and down, slamming down on hard ground, you're seeing stars… When you get off and try to set up, your body temperature has gone so low that as you touch anything you get incredibly cold and your hands barely function as you try to film with a frozen viewfinder and a camera that doesn't work. So it was quite an experience, but not one I'd ever want to repeat.' By comparison, bumping along on a Ski-Doo in the Arctic tundra to film Arctic wolves was also incredibly painful, but 'on the other hand,' says Mark, 'travelling with a pack of wolves in the middle of a hunt is the most incredible experience you can have.'

BIG PREDATORS AND SERIAL BURGLARS

When filming the trials of nesting Arctic birds and their predators, assistant producer Sophie Lanfear and cameraman Barrie Britton had the comparative luxury of a summer shoot on a remote Svalbard island. Their accommodation was a one-room, 4-metre-square miners' cabin, which they shared with a Norwegian guide and 25 cases of equipment.

Though there was 24-hour daylight, they had just four days of sun out of six weeks of rain and fog. But they did have characterful birds and mammals. On the unlikely chance of a polar bear visit, they had a flare gun each and a rifle for the hut. But for five weeks, the only visitor to the cabin was Slikkerpott (Norwegian for pot-licker), the Arctic fox, and the only real trial was running the gauntlet of dive-bombing nesting skuas and gulls on the 400-metre approach to the beach toilet. It wasn't until week six, when the guillemot chicks were starting to jump and filming was at its busiest, that a troublesome visitor arrived on the island.

◀ **Home for the Svalbard crew.** The old miners' hut where the crew lived for seven weeks. When weather closed in, which it did for half the time, the crew would be confined to the one room.

▼ **The pot-licker.** Slikkerpott, the young fox who would turn up every evening, smelling cooking. He would curl up outside and wait, looking up pleadingly if anyone emerged, and would do the washing up if the pots were left out.

They first spotted the polar bear when they were on their way back from the cliffs for a few hours' sleep. As they reached the cabin, they could see that the door was open. 'He'd pulled it off its hinges, gone through all the shelves and eaten everything, including my two bars of Toblerone,' says Sophie. He'd also managed to open the locked freezer outside – '26 kilos of meat, cheese, yogurt – gone in one sweep'. It took three hours for them to clear up the mess and barricade the door, only to lie awake in anticipation of the bear's return. Sure enough, at 4am there was a pounding at the door. They scared him off by shouting, screaming and banging the walls – but not for long.

The next day, from the cliffs, they spotted him heading for the cabin. At that point, there was nothing they could do. Returning at midnight, they found him resting near the cabin, red-wine stains down his face. 'This time he had really gone to town,' says Sophie. The bear-proof door was in pieces, and he had torn open everything – even the wine boxes. 'It was worse than a teenage house party… He had smashed all the jars, cleared the shelves,

◀ **After the break-in.** The hut in the process of bear-damage repair, the door back on its hinges but the bear-guard propped up to show the ease in which it had been ripped off. It was impossible to mend it as there was no wood left.

▲ **Burgled meat store.** Remnants of the contents of the freezer-chest. The bear ate all the meat – 20kg of it – and the dairy items, skilfully sucking out the packet contents and delicately licking out the yogurt pots.

▲ **After the party.** Evidence of the polar bear feast, including chocolate wrappers and the remains of the red wine box. The perpetrator was spotted near the hut, asleep, his muzzle stained with wine.

▶ **Strength of evidence.** The storage area, revealing the door ripped to pieces and anything edible inside eaten.

opened drawers and looked into cupboards – incredibly dexterous really. There was polar bear hair and slobber over everything… The only stuff he didn't eat was the gin, the Marmite and the washing up liquid' – and, luckily, the tinned food and pasta in the attic. 'Everything else was all over the floor in one congealed glob.' That clean-up took about four hours.

There was no option now but to mount a 24-hour watch. Sophie stayed up that night with the gun beside her, while the other two slept, so they could go filming in the morning. 'I knew how clever bears could be and knew he would be calculating what to do next.'

Sure enough, after the others had left, he appeared at the window. Screaming scared him off again, and subsequently explosives worked. But the toilet run to the beach remained nerve-wracking, remembers Barrie.

On their return to Bristol, they learned that the bear was indeed a serial burglar – a specialist in cabin break-ins – and had finally broken into the cabin again, this time smashing through the window.

RAINFOREST PERILS

For Jonnie Hughes on the Ecuador shoots, bites were a constant problem – 'soldier ants were especially painful' – with a risk, especially when filming at night, of bites from vipers and venomous insects. But a greater danger was getting lost in the jungle: 'It's so easy to turn around without realizing and head off in the wrong direction.' In the end, they tied ribbons along the routes to make sure they could find their way back to camp.

For Adrian Seymour, filming harpy eagles in Venezuela, the fascination was how successfully the eagles blended into the rainforest. 'You never see or hear one coming. You might see it to start with, then it disappears out of vision. Just when you think it's gone, it hits you... One time I was up in a tree 250 metres away from the nest, and she hit me in the back with all her talons, and all of them went in.'

Of course, the eagle was just defending her youngster from a perceived threat. And the team took great care to build the scaffolding at a distance and only to climb up it to film when they thought the mother harpy was away from her nest. As it turned out, the greatest danger was the climb itself.

Producer Rob Sullivan was there to film the harpy eagle scientist Alexander Blanco at work for the last programme. Dr Blanco has been studying the eagles for 20 years, ringing and putting transmitters on them to monitor how they are coping with the fragmentation of their habitat. On this occasion, he had climbed to the nest, where he was interviewed by Adrian, and had then wrapped up the huge chick to carry it down to be ringed. When he stepped off the branch and leant back to take his weight on the rope, the anchor point failed and the rope slipped straight through the pulley.

'He just disappeared into the canopy – fell 100 feet,' says Rob. I was supposed to be filming him abseiling. Instead I was filming him falling. I could hear him screaming in my earphones all the way down. My first thought was that he surely must be dead.'

Miraculously, he wasn't. But he had badly broken his wrist and femur and was in agony. The team made a stretcher out of sticks and a poncho, carried him a mile to the camp and then drove five hours on bumpy roads before

▶ **Hot work.** The female harpy eagle rests on her nest high up in the canopy, panting from the heat. She is being filmed from 40 metres (130 feet) away. Filming was helped by the fact she was used to the presence of researchers in the forest below.

◀ **Chick monitor.** Dr Alexander Blanco's research assistant Don Blas (left), helped by producer Adrian Seymour, gently attach a transmitter to the five-month-old harpy eagle chick before returning it to its nest high in the nearby kapok tree on the edge of a forest clearing.

▶ **Top view.** Seated all day on top of a 30-metre-high (100-foot) scaffolding tower in a clearing opposite the nest tree, cameraman Rob Sullivan films the harpy eagle activities. The parents were so used to the researchers that no hide was needed. Matt Aeberhard spent four weeks sitting on the same platform filming the chick and another four filming it when it had fledged and was learning to hunt. John Aitchison spent five weeks filming the fledging chick and its parents from a tree platform within the forest.

reaching somewhere that had morphine. When they finally got to a hospital, Dr Blanco insisted that the team carry on filming – so the world would see his beloved birds and the problems they face.

Miraculously, the chick was unharmed and placed back in the nest with a transmitter attached. Alexander couldn't walk for four months – and couldn't teach at the university, which meant no income for his research (he funds this himself, as there is little money for conservation in Venezuela). But he is now back on the project, tracking the chick and its progress.

In the field, there is always the danger of accident or illness, but on *The Hunt*, there was seldom real danger from the predators. Says Jonnie Hughes of the Arctic wolves, 'I saw how vicious they can get [having seen them kill an intruder from another pack], and they could have killed one of us had they really wanted to. But I always felt safe.'

The marine shoots were among the most dangerous, not because of sharks but the possibility of losing one of the team in the ocean or, more likely, being run over by the boat. But rigorous risk assessments and a highly experienced team made sure such events didn't happen. Certainly, this was one series where the team returned safe and triumphant.

INDEX

Page numbers in **bold** refer to the illustrations

A

aardvark 100

Abdopus octopus 140, **140**

adder, puff **98–9**, 99

Aeberhard, Matt 257–8, 303

Africa *see individual countries and animals*

African wild dog 9, 21, 25–9, **26–9**, 242–4, **242–3**, 264, 268

Aitchison, John 268, 303

Alaska 43, 45, 48–9, 145–8

albatrosses 251

   black-browed **216–17**

   grey-headed 218

   wandering 216–18, **219**

   waved 218

Amazon 70, 252

ambushes

   harpy eagle 70

   Portia jumping spiders 70–3

   sparrowhawk **2–3**, 58, **60–1**

   tiger 65

amphipods **224**, 225

ampullae of Lorenzini 207

Amur falcon **34–7**, 37

anchovies 207

Anderson, Doug 254, 278–82, **280**, 290, 293

anglerfishes 229–30

Antarctica 46, 49, 153, 162, 168, 172, 218, 233

antbirds 82

anteater, giant 100

antelopes 18–21, 86, 93

   duikers 21, 22

ants

   army 78–83, **78–83**, 252, **253**

   Azteca 78–9

   hotrod 122–4, **122–3**

   soldier 300

Arctic 162–8, 171–97

   birds 106, 172–6, 179–80, 184, 194–7, 271

   killer whale 43

   muskox 187–8

   polar bear 38–40, 165–8, 171, 182–4, 190–2, 197

   seals 153, 166, 171, 182–3, 192 *see also individual species*

   walrus 197

   whales 233

Arctic fox **160–1**, **295**, **297**

   coat 171, 178

   diet of birds and eggs 172, 175, 176, 179, 180, **180**, 194–6, **273**

   filming 271–2, 294–5, 297

   and polar bear **170–1**, 171

Arctic hare **162–3**, **176–9**, 179, 187, **187**, 244, 284–6, **286**

Arctic tern 175

Arctic wolf 178–9, **186–9**, 187–8, 244, 268, 284–6, 295, 303

Argentina 157–8

Atkinson, Ryan **274–5**

Atlantic Ocean 208, 215, 218

auk, little 172–5, **175**, 184, 271–2, **272–3**

Australia 46, 137–40, 152

B

baboons 259

badger, honey 96–9, **97–9**

baitballs 209, 210, **215**, 278, **280**

Bale Mountains 113

Bandhavgarh National Park 52, **240–1**, 244–7

barnacles 142

Barnett, Luke 252, 290

Barren Island **166–7**

beaches 137–40, 148–51, 152–4

bears

   brown 70, **126–7**, **144–9**, 145–8

   grizzly 148, 165

   see also polar bear

Beaufort Sea 168

Bedwell, Andy **251**

beetles 70

   click (headlight) **100–3**, 102–3

Bengal tiger 62–6

Bering Sea 43, 45

beroes 205

billfishes 17, 207, 208–9, 210, 222, 278

bioluminescence **101–3**, 226, 229–30

birds

   in Arctic 106, 172–6, 179–80, 184, 194–7, 271

   migration 33–7

   waders 17, 131–4, 175, 176

   see also individual species

birds of paradise 52

Blanco, Alexander 300–3

Blas, Don **302**

blue whale **202–3**, **232**, 233–5, **234–5**, 238, **238–9**, 261, **262–3**

Brazil 102

breathing, cheetah 21

Britain 131–4

Britton, Barrie **270–1**, 271–2, 297, 299

brown bear 70, **126–7**, **144–9**, 145–8

buffalo **14–15**, 24, 116–21, **116–21**

bushbuck 21

buzzard, steppe 37

C

California 43, 45, 148, 261

cameras 9

  camera dollies **275, 276**

  Cineflex 241–51, **246, 258,** 259

  infrared 257, **257**

  mini-cameras 252, **253**

  polecam 254

  slow-motion 268

camouflage 30, 52, **62-3**

  Abdopus octopuses 140

  in Arctic 175–9, 180

  caracal 94

  leopard 21

  in oceans 226, 229

  Portia jumping spiders 73

  tigers 63–5

Canada 14, 109–10

capelin 148–51, **150-1**

caracal **92–5,** 93–4

Caribbean 151

caribou 14, 176

Central Africa 14, 22

Central America 78

cheetah **6–7, 18–19, 84–5, 88–91,** 96, **96, 265**

  breathing and heart rate 21

  filming 238, 264

  hunting strategy 18–21

  hunting success 19–21, 23

  life expectancy 90

  loss of kills 89–90

  speed 18–21, 89

Chile 267

chimpanzee 22, 76–7, **76–7**

China 37

Cineflex cameras 241–51, **246, 258,** 259

click (headlight) beetles **100–3,** 102–3

cliffs 172–3, **184–5,** 194–6, **270–1,** 271

coasts 128–59

cod 151

colonies

  army ants 78

  seabirds 172, 173–5

  seals 171

  snow geese 179, 184

  termites 100–3

  walruses 197

comb jellies **204,** 205

cooperative hunting *see* social predators

Cordey, Huw 238, 242, 244, **245, 258,** 259, 264, 274–7, **274, 275, 288–9,** 294

Costa Rica 152, 255

counter-shading, fish 207

coyotes 58

Crab Island 152

crabs 142

  sand bubbler 137, 138, **138–9**

  soldier **136–7,** 137

crocodiles, saltwater 152

crows 142

Crozet Island 157, 218

ctenophores **204,** 205

curlew, long-billed 131

cutlass fish 226

D

dark zone, oceans 229–30

Darlington, Sophie **8,** 267, 294

Darwin's bark spider 30–3, **30–1,** 274–7

deer, chital 62–3, **64–5,** 247

deserts 122–4 , 154-5

Dimbleby, Robin **244**

Dolch, Rainer **275,** 277

dolphins 210, 221

spinner 210, **210–11,** 254–5, **254–5**

dorado 221

Doyang reservoir, Nagaland **36**

dragonfishes 230, **230**

ducks 17

  eider 175–6, 184, 194

duikers 21, 22

E

eagles

  bald 106–7, **106–7,** 151

  harpy 9, 66–70, **67–9,** 300–3, **300–3**

ears *see* hearing

echolocation 210, 233

Eciton burchellii 78, **80–1,** 82–3

Ecuador 70, 252, 300

eels

  gulper 229

  snipe 226, **227**

eider ducks 175–6, 184, 194

eland 21

elephants 24, **240,** 245–7, **246**

Ellesmere Island **162–3, 176–7,** 180, **187, 188–9,** 284–6, **284–7**

Ethiopian wolf **112,** 113–15, **114–15,** 294

Etosha National Park 25, 124–5, **124–5, 266–7,** 267

eyesight

  fish 208, 225

  harpy eagle 69

  tarsiers 73

  tigers 65

F

falcons

  Amur **34–7,** 37

  peregrine **16–17,** 17, 134, **134–5**

Falkland Islands 157

fish 203
  counter-shading 207
  eyesight 208, 225
  lateral lines 207
  migration **32**, 33, 208
  shoals 18, 200, 207, 215, 278–81
  speed 208
  see also individual species

flocks
  snow goose 103–7
  waders 134

Florida **32–3**, 33

flying fishes 221

Fothergill, Alastair 241, 260–1

foxes 58, 158
  red 150
  see also Arctic foxes

frigatebirds **198–9**, **220**, 221, 251

G

Galapagos Islands 218

gannet 218

gaur 62

gazelles 264
  Grant's **90**
  Thomson's 10–11, **18–19**, 19–21,
    84–5, 90

geese
  greater white-fronted **108–9**,
    109–10
  pink-footed 131, 132
  snow 86, 103–7, **104–7**, 179–80,
    180–1, 184

ghostly seadevil **228**

giant mole rat 113–15, **113**

glass squid **228**

godwits 131

goshawk, pale chanting 99

Greenland 38, **173**

group hunting *see* social predators

grunion 148

guillemots 172, **172–4**, 184, 194–6,
  194–5, **270–1**, 271, 297

guineafowl 94, **94–5**

gulls 142, 297
  glaucous 172, 173, 175, 196, 271
  great black-backed **195**

gulper eel 229

gyrfalcon 172, **173–5**

H

halibut 151

hare, Arctic **162–3**, **176–9**, 179, 187,
  187, 244, 284–6, **286**

harpy eagle 9, 66–70, **67–9**, 300–3,
  300–3

hatchetfishes 225–6

Hawaii 210

hearing
  caracal 93
  harpy eagle 69
  tarsiers 74
  tigers 65

heart rate, cheetah 21

heat tolerance 122–4

helicopters 244, **244–5**, 264, 290

Hellevik, Jostein 268

herds
  Arctic hare 179
  buffalo 116–17

herons 137

herrings **42–3**, 151, 207, 215

hibernation 40

Himalayas 37

honey badger 96–9, **97–9**

Hughes, Jonnie 245, 247, **249**, 252, **253**,
  285, 300, 303

humpback whale 46, **46–7**, 151, 233,
  238, 251, 261, 289–93, **290–2**

Husain, Ellen 238, **288–9**, 289–93, **293**

hyenas 19, 25, **28–9**, 89
  brown 154, **154–5**
  spotted 23, 29, **91**, **257**

I

impala 18, 21, 258, 259

India 37, 52, 62–6, 244–7

Indonesia **75**, 140

infrared cameras 257, **257**

Iran 93

Ivory Coast 76–7

J

jackal, black-backed **155**

jaguar 83

jay, Eurasian **2–3**, **60–1**, 268, **268–9**

jellyfish 204–5, **231**

K

Kalahari Desert 99

Katmai National Park 145–8

Kaziranga National Park **62–3**

killer whale 9, 41–9, **42–9**, **156–9**, 157–8,
  168, 238, 251, 261, 289–93, **290–1**

kingfishers 137

kittiwakes 172

knot **132–5**, 134

krill **202**, 203, 218, 233–5, 261

Kruger National Park 23

L

Lanfear, Sophie 266, **270–1**, 271–2,
  297–8

langur 62–3

lanternfish 226, **227**

larvaceans 204

lateral lines, fish 207

leaping, caracal 93

lemmings 110, **110–11**, 180, 187, 188

leopard **1**, 14, 19, **20–3**, 21–3, **256–7**, 257–9, **259**, 294

life expectancy, cheetah 90

light
bioluminescence **101–3**, 226, 229–30
photophores 226, 230, **230**

limpets 142

lion 9, **10–11**, **14–15**, 19, **266–7**
and African wild dog 29
filming 267
hunting buffaloes 116–21, **116–21**
prides 24–5, **24–5**, 124–5, **124–5**
stealing kills 21, 23, 89

Liuwa Plain National Park 257, 264

lobster 142

Luangwa National Park 117

lunge-feeding, whales 151, 234–5

M

macaque, long-tailed **4–5**, 8, 142, **142–3**

MacEwen, Alastair 252, **253**, **274–5**

mackerel, Indian **206–7**, 207

McPherson, Jamie **240–1**, 241–7, **244**, **246**, **249**, **251**, 258, 259, 264, 286

Madagascar 30–3, 274–7

Mallalieu, Ilaira **253**

marlin 204
striped **208–9**, 209, 278

marten, American pine **54–7**, 55–8

Massai Mara 86

Mexico 33, 45

migration
birds of prey 33–7
fish 33, 208
in oceans 226
sharks 33
snow geese **103–7**
whales 45, 233, 289
wildebeest 86

mimicry 30, 73

mole rat, giant 113–15, **113**

Molteno, Frank **244**

Mongolia 37

Monk, Blair **288–9**, 290, **293**

monkeys 21, 22
howler 69–70
long-tailed macaque **8**, 142, **142–3**
red colobus 76–7
tamarin 82

Monterey Bay 45

mudflats 131–4

muskox 187–8, **188–9**, 244, **285**, 286

mussels 142

N

Nagaland **36**

Namib Desert 122–4, **122–3**

Namibia 23, 124–5, 154

nektonic animals (nekton) 203, 207

Nepal 37

Newfoundland 148–51

night heron 152

Ningaloo Reef 289, **290–1**

nocturnal animals 22–3, 73–4

Norway **42**

O

oceans 198–235

octopuses **128–9**, 225
Abdopus 140, **140**

orca *see killer whale*

orb webs, Darwin's bark spider **30–1**, 30–3, 274

otter, marine **128–9**, 141, **141**, 267, 281–2, **282–3**

owl, snowy **108–11**, 109–10, 180

oystercatchers 142

oysters 142

P

Pacific Ocean 42

Palearctic 37

peafowl 21

Pearson, Hugh 238, 255, 261, 266, 278–81, 282

pelicans 152

penguins 172

Peninsula Valdes 157–8

peregrine **16–17**, 17, 134, **134–5**

phalaropes 176

Philippines 140

photophores 226, 230, **230**

phytoplankton 144, 203, 218, 221

pigeons 17, 93

pine marten, American **54–7**, 55–8

Pitman, Bob 289

plankton 144, 203–7, **204–5**, 226, 261

plovers 131

polar bear 9, **38–41**, 153, **164–5**, **168–71**, 182–3, 190–2, **196–7**, 250
Arctic fox and **170–1**, 171
coat 165
cubs 182–3
filming 248–51, 266

home range 38
hunting strategies 38–40, 166–8, 183, 192, **192–3**, 197
physiology 40
raiding bird colonies 172, 175, 184, **184–5**
raiding cabin 297–9, **298–9**
swimming 190–1, **190–1**
polar regions *see* Antarctica; Arctic
polecam 254
porcupines 21
Portia jumping spiders 70–3, **70–2**
Portuguese man-o'-war 205
Pribilof Islands 45
pride, lion 24–5, **24–5**, 124–5, **124–5**
ptarmigan 188
pteropod snails 204, **204**

R
radio collars 268
rainforests 300
  army ants 78–83
  camouflage 52
  filming in 252, 300–3
  harpy eagle 9, 66–70, 300
  spiders 30–3, 70–3
ratel *see honey badger*
rays 207
red river hog 22
redshank 134
Reichert, David **238–9**, 261, **262–3**, 278, **278–9**
Russia 37

S
sailfish 17–18, 208, **220**, 222
salmon
  chinook 42

sockeye 145–8, **148–9**
saltpans 124
sand-hoppers 134
sardines **200–1**, 207, **208–9**, **212–15**, **220**, 261, 278–81, **278–80**
Scholey, Oliver 295
sea gooseberries 205
sea spiders 226
seadevil, ghostly **228**
sealion 152, 157–8, **158–9**, **212–15**, 278, **278–9**
seals 152–3
  bearded 40, 171, 191–2, **192–3**
  crabeater 49
  elephant 154, 157, 222
  fur 45, 154, **154**
  harp 168, 171
  hooded 171
  leopard 153, 168
  ringed 38, **38**, 166–8, **166–7**, **169**, 171, 182–3, **182–3**, 190
  Weddell **48–9**, 49, 168
Serengeti 14, 21, 23, 25, 86
Seymour, Adrian 268, 300, **302**
sharks 207, 208
  black-tip **32–3**, 33
  blue 222, **222–3**
  bull 33
  copper 278
  hammerhead 33, 207
  Pacific sleeper 49
  shortfin mako 17
  spinner **32–3**, 33
  whale 207
shearwaters **278–9**
shoals, fish 18, 33, 200, 207, 215, 278–81
silk, Darwin's bark spider 30–3, **30–1**, 274

Simien Mountains 113
siphonophores 205, **205**, **231**
skuas 175, 297
sloths 69
smell, sense of
  albatrosses 218
  honey badger 99
  polar bear 166
Smith, Mark 244, 267, 281, **281**, 284–5, **284–6**, 294–5
snails, pteropod 204, **204**
snakes
  puff adder **98–9**, 99
  snipe eel 226, **227**
  snow 56–8, 295
  soaring 217–18, 221
social predators
  African wild dog 25–9, **26–9**
  chimpanzees 76–7
  dolphins 210, **210–11**
  killer whale 42–9, 157–8
  lion 24–5, **24–5**, 124–5
South Africa 23, 37
South America 33, 78, 141, 281
South Georgia 153–4, 218
Southern Ocean 49, 216–18, 233
sparrowhawks **2–3**, 58–9, **58–61**, 66, 134, 268, **268–9**
speed 17–21
  cheetah 18–21, 89
  fish 208
  leopard 21
  lion 25
  sparrowhawk, Eurasian 58
  tigers 65
spiders 9
  Darwin's bark 30–3, **30–1**, 274–7
  decoy 277, **277**

jumping 70–3, **70–2**

spitting **70–1**, 73

spoor **123**, 124

spinner dolphin 210, **210–11**, 254–5, **254–5**

springbok 18

Spitzbergen *see Svalbard*

Squaw Creek, Missouri 103–7

squid **224**, 225

big-fin **228**

glass **228**

stalking

caracal 94

cheetah 19

Ethiopian wolf 114

harpy eagle 70

leopard 21–3

polar bear 183, 190–2

tigers 65

Stark, Mandi **245**, 257–8, **258**, 267, 294

steppe buzzard 37

stick insects 79

still-hunting, polar bear 183

stoop, peregrine falcon 17

stranding, killer whale **156–9**, 157–8

Sulawesi **75**

Sullivan, Rob 300, **303**

sunfish 203

Svalbard 172, **174**, **184–5**, **196–7**, 266, 271, **296–9**, 297–9

swordfish 208

T

Tai forest, Ivory Coast 76–7

tarantulas 78

tarsiers 73–4, **74–5**

teeth, killer whale 48

temperatures

in deep ocean 222

extreme 294–5

heat tolerance 122–4

termites 37, 100–3, **100–3**

tern, Arctic 175

territories, tiger 65

Thailand 142

Tibetan Plateau 37

tiger, Bengal **50–1**, 62–6, **62–5**, 244–7, **247**

tool use **4–5**, 142, **142–3**

toothfish, Antarctic 49

Totterdell, John 289

traps, Darwin's bark spider 30–3, **30–1**

tuna 204, 207, 208, 210, 221, 278

bluefin 208

yellowfin 208

tundra 14, 109–10, 172, 176–9

turtles

flat-backed 152

leatherback 222

olive ridley 152, **152–3**

twilight zone, oceans 225–6

U

ultrasound 74

Unimak Pass 45

V

Venezuela 300–3

Veracruz 33

viperfish **225**

voles 55–8

vultures 89

W

waders 17, 131–4, 175, 176

wahoo 208

walrus 40, 152, **196–7**, 197, 266

The Wash **130–1**, 131–3

waterholes 117, 124, **243**

webs, Darwin's bark spider 30–3, **30–1**, 274–7

Weddell seal **48–9**, 49, 168

whales 14, 203

blue **202–3**, **232**, 233–5, **234–5**, 238, **238–9**, 261, **262–3**

Bryde's **215**, **280**, 281

grey **44–5**, 45

humpback 46, **46–7**, 151, 233, 238, 251, 261, 289–93, **290–2**

minke 49

sperm 207

wild dog, African 9, 21, 25–9, **26–9**, 242–4, **242–3**, 264, 268

wildebeest 9, **24–7**, 28–9, 86, **86–7**, 117, **243**, 264, **265**

willet **16–17**

wolves 14, **147**, 148, 284–6, **287**

Arctic 178–9, **186–9**, 187–8, **244**, 268, 295, 303

Ethiopian **112**, 113–15, **114–15**, 294

Wrangel Island **160–1**, 180

Z

Zambia 117, 242–4, 257, 264, 294

zebras 267

ACKNOWLEDGEMENTS

10 9 8 7 6 5 4 3 2 1

BBC Books, an imprint of Ebury Publishing
20 Vauxhall Bridge Road,
London SW1V 2SA

BBC Books is part of the Penguin Random House group of companies, whose addresses can be found at global.penguinrandomhouse.com

Copyright © Alastair Fothergill, Huw Cordey 2015

Alastair Fothergill and Huw Cordey have asserted their right to be identified as the authors of this Work in accordance with the Copyright, Designs and Patents Act 1988

This book is published to accompany the television series entitled *The Hunt*, first broadcast on BBC1 in 2015

Executive producer: Alastair Fothergill
Series producer: Huw Cordey
BBC Head of Commissioning, Natural History: Tom McDonald
BBC commissioning editor: Kim Shillinglaw

First published by BBC Books in 2015

www.eburypublishing.co.uk

A CIP catalogue record for this book is available from the British Library

ISBN 978 1 84990 722 4

Commissioning editor: Albert DePetrillo
Project editor: Rosamund Kidman Cox
Editor: Kate Fox
Designer: O'Leary & Cooper
Picture researcher: Laura Barwick
Image grading: Stephen Johnson, Copyrightimage
Production: Antony Heller and Alex Goddard

Colour Origination by AltaImage, London
Printed and bound in Italy by LEGO

Penguin Random House is committed to a sustainable future for our business, our readers and our planet. This book is made from Forest Stewardship Council® certified paper.

This book and the television series it accompanies aim to look at the dynamic relationship between predators and prey in a fresh and detailed way. We hope to dispel something of the negative reputation predators have and reveal them for what they truly are – the most impressive and hardest-working animals of all. But filming hunting behaviour is not easy. Hunts are rare and predators are unpredictable, and capturing the key moments of the chase means being in the right place at exactly the right time. Happily, we were to benefit from the wealth of experience and knowledge generously offered by scientists and field assistants from all around the world. We hope we have included all their names in the long list of thanks and remain very grateful for everything they did to make the series and this book possible.

The responsibility for capturing all the fleeting moments of exciting action lay in the hands of our determined and talented team of camera operators. We are particularly grateful to the principal team of just five who filmed more than 50 per cent of the series.

The extraordinary production team based at Silverback Films in Bristol worked tirelessly for three years to bring so many fresh stories back from the field. And this would have been totally impossible without the support of the best production management team in the business. We were very fortunate that an outstanding postproduction team worked their special alchemy to finish the films in the best possible way.

We are very grateful to Steven Price for his powerful and original score. And it was once again a privilege that David Attenborough narrated the series with the clarity and poetic enthusiasm that only his voice can bring. Huw and I would also like to thank Albert DePetrillo, for commissioning this book, our agent Sheila Ableman for all her kind support and our picture editor Laura Barwick, our designer Tara O'Leary and our editor Rosamund Kidman Cox (who also wrote chapter 7) for their relentless determination to ensure the book was the very best it could be – only the hunting dogs of Zambia compare with them in terms of energy and stamina.

PRODUCTION TEAM
Dan Clamp
Darren Clementson
Jenni Collie
Rebecca Coombs
Huw Cordey
Marcus Coyle
Charles Dyer
Sarah Edwards
Alastair Fothergill
Jane Hamlin
Hal Hampson
Jonnie Hughes
Ellen Husain
Rachel James
Tara Knowles
Sophie Lanfear
Ilaira Mallalieu
Katie Mayhew
Elisabeth Oakham
Alex Page
Hugh Pearson
Sarah Pimblett
Jason Roberts
Adrian Seymour
Vicky Singer
Hannah Smith
Mandi Stark
Rob Sullivan
Rose Wilson

PRINCIPAL CAMERA TEAM
John Aitchison
Doug Anderson
Sophie Darlington
Jamie McPherson
Mark Smith

CAMERA TEAM
Matt Aeberhard
Luke Barnett
Malcolm Beard
Tom Beldam
Barrie Britton
Richard Burton
Rod Clarke
Darren Clementson
Robin Cox
Mark Deeble
Stephen De Vere
Kevin Flay
Nick Guy
Graham Hatherley
Jonathan Jones
Michael Kelem
Simon King
Ian McCarthy
Alastair MacEwen
Robert McIntosh
Blair Monk

Peter Nearhos
Didier Noirot
Kieran O'Donovan
Mark Payne-Gill
David Reichert
John Shier
Warwick Sloss
Robin Smith
Rolf Steinmann
Vicky Stone
Bali Strickland
Rob Sullivan
Gavin Thurston
Jesse Wilkinson

FIELD ASSISTANTS
Ryan Atkinson
Andy Bedwell
Kira Cassidy
Heather Chambers
John Chambers
Corinne Chevalier
Jacca Deeble
Robin Dimbleby
Einar Eliassen
Håvard Festø
Pennie Ginn
Dean Miller
Robert Myler
Alonso Sanchez

Oliver Scholey
Oliver Saurabh Sinclair
Charlie Stoddart
Oskar Strøm
Gisle Sverdrup
Audun Tholfsen
Ben Tutton
Ignacio Walker

POST PRODUCTION
Films at 59
Bridget Blythe
Gordon Leicester
George Panayiotou

MUSIC
Steven Price
BBC Concert Orchestra

FILM EDITORS
Nigel Buck
Tim Lovell
Matt Meech
Andy Netley
Dave Pearce
Sam Rogers

ONLINE EDITORS
Simon Bland
Franz Ketterer

DUBBING EDITORS
Kate Hopkins
Tim Owens

DUBBING MIXER
Graham Wild

COLOURIST
Adam Inglis

GRAPHIC DESIGN
Burrell Durrant Hifle

**THE OPEN
UNIVERSITY**
Miranda Dyson
Caroline Green
Caroline Ogilvie
Janet Sumner
Vicky Taylor

More information about
The Hunt from:
www.open.edu/
openlearn/thehunt

SPECIAL THANKS TO
Aerial Filmworks
African Parks Zambia
Alaska's Hallo Bay
  Wilderness Camp
James Aldred
Kirsty Allen
Andasibe-Mantadia
  National Park
Charles Anderson
Morgan Anderson
Arctic Institute of North
  America
Adam Ashraf
Association Mitsinjo
Katy Austin
Avitrek
Bandhavgarh National
  Park
Andrés Vallejos Baier
Balai Konservasi Sumber
  daya Alam (BKSDA),
  Tangkoko,
  North Sulawesi
Robert Bartlett
Peter Bassett
Christine Bays
Gerard Beaton
Janice & Richard Beatty
Matt Becker
Eric Bedin
Colleen Begg
Ronald L .Bell

Maristela Benites
Abhra Bhattacharya
Bioparque El Puquen
Dave Blackham
Alexander Blanco
Don Blas
Reyk Boerner
Andre Botha
Espen Brandal
Paul Brehem
Femke Broekhuis
Peter Brownlee
Elisabeth Brox
Bureau of Protected
  Areas, Chubut
John Calambokidis
Jane Carter
Hector Casin
Centro Nacional
  Autónomo de
  Cinematografía
  (CNAC)
Costa Cetacea
Rob Clifford
Ella Cole
Octavio Colson
Reginaldo Constatino
Javier Contreras
Juan Copello
Ricardo Correa
Cleide Costa
Corporación Nacional
  Forestal
Martin Cray
Will Cresswell
Antica Culina
Marcos Da Silva Cunha
Charlotte Demers
Department of National
  Parks, Thailand
Thoswan Devakul
Robin Dimbleby
Rainer Dolch
Pastora Donoso
Tom Doyle
Egil Dröge
Edriss Ebu
Emas National Park
Estación de Biodiverisdad
  Tiputini
Ethiopian Wildlife
  Conservation
  Authority
Ethiopian Wolf
  Conservation
  Programme
Etosha National Park
Patrick Evans
Chris Evans

Edmund Farmer
Lynn Faust
Ola Fincke
Julian Finn
Joshua Firth
Nigel Fisher
Dan Fitzgerald
Tom Foreman
Adam Fox
Gates Underwater
  Products
Al Gaudet
Diane Gendron
Weldy George
Graeme Gillespie
Great Smoky Mountains
  National Park
Gobabeb Training and
  Research Facility
Lance Goodwin
Matjaz Gregoric
Michael David Gumert
Steve Haddock
Bano Haralu
Nagruk Harcharek
Jean Hartley
Ibrahim Hassan
Peter Hawkes
Emily Haynes
Will Hayward
Iostein Hellevik
Richard Herrmann
Jeff Hester
Jimmi Hill
Denver Holt
Danny Howard
José Rafael Hurtado
  'Cheo'
John Innis
Ingela Jansson
Zoe Jeffery
Paul Jensen
Martin Jørgensen
Paul Johnsgard
Kornelius Jonas
Ullas Karanth
Krithi K. Karanth
Natasha Karniski
Katmai National Park,
  National Park Service,
  US Department
  of the Interior
Roland Kays
Anna Keeling
Kimberley Marine
  Research Station
Kluane National Park and
  Reserve
Ree Komatsu

Michael Kristjanson
Kruger National Park
Meemendra Kumar
Bjørne Kvernmo
Roimen Lelya Laizer
Ramnaresh Barman Lala
Bobby Lambaihang
Stephen Lang
Nathalie LaSalle
Don Lavallee
Pamela Lepe
Daiqin Li
Mark Linfield
Liuwa Plain National
  Park
LookBermuda
Moloimet Kilusu
Lukumay
Nick Lunn
Maasai Mara National
  Reserve
Ben MacDonald
Jodi MacGregor
John MacIver
Yadvinder Malhi
Kolei Ikayo Mamasita
Kevan Mantell
John Marchant
Juan Marín
Marine Studios & Florida
  Biodiversity Institute
Colette Massier
Malcolm McAdie
Tim McCagherty
Lorraine and Bob McGill
Dan McNulty
Stacy McNulty
Gonzalo Medina-Vogel
Andres Emilio
  Perez Mejias
Javier Mesa
Jeffery S. Mesach
Monterey Bay Aquarium
  Research Institute
Sammy Munene
Phillimon Mwanza
N/a'an ku sê
Nagarhole National Park
Brian Nakashima
Namib-Naukluft Park
Amit Nayyer
The New Island
  Conservation Trust
Ngorongoro
  Conservation Area
  Authority
Carey Nicholson
Letro Nini
Roger Niño

Sean O'Donnell
Steve Oliver
Owl Research Institute
Craig Packer
Pangti Village
Parks and Natural Areas
  Division,
  Newfoundland &
  Labrador Department
  of Environment
  & Conservation –
  Mistaken Point
  Ecological Reserve
Parque Nacional Pan
  de Azúcar
Ange Peers
Margie Peixoto
Jack Pettersen
Ian Phillips
Rita Pikasi
Nathan Pilcher
Robert Pitman
Simon Pitt
Polar Continental Shelf
  Program, Natural
  Resources Canada
Simon Pollard
Jerome Poncet
Leith Poncet
Koos Potgieter
Rina Pretorius
Qikiqtani Inuit
  Association
Jaime Quispe Nina
Paul Ratson
Jenny Read
Lary Reeves
Reserva Nacional
  Tambopata
Claire Revekant
Evan Richardson
Neon Rio
Terhi Riutta
RSPB Snettisham
Barry St George
Ryan St John
Paul Saroli
Conway Sassoon
Keith Scholey
Angela Schuler
  Brennan
David Seaman
Secretariat of
  Environment and
  Natural Resources
  (SEMARNAT)
Émilie Sénécal
Serengeti National Park
Ben Sheldon

Salamonie Shoo
Claudio Sillero
Leandro Silveira
Toby Sinclair
Digpal Singh
Daan Smit
Diana Smith
Dylan Smith
Jonathan Smith
Theresa Smith
Zak Smith
Scuba Travel
Smithsonian Institution
South Luangwa National
  Park
Squaw Creek National
  Wildlife Refuge
Debbie & Rick Stanley
Ian Stirling
Tambopata Research
  Centre
Tanganyika Film and
  Safari Outfitters
Daphne Taylor
Chris Timmins
Phil Timpany
Phil Torres
John Totterdell
Tswalu Kalahari
Umiaq
US Fish and Wildlife
  Service
Derek & Claire van der
  Merwe
Vincent van der Merwe
Gus van Dyk
Rudie van Vuuren
Marlice van Vuuren
Viewfinders
Vadim Viviani
Wapusk National Park,
  Parks Canada Agency
Stefanie Watkins
Tim Watson
Darrin Welchert
Linda Weldon
Coli Whelen
Niklas Wikstrand
Wildlife Conservation
  Society
Jackie Willis
Greg Willis
World Bird Sanctuary
Mark Young
Zambian Carnivore
  Programme

PICTURE CREDITS

1 Federico Veronesi; 2–3 Pål Hermansen; 4–5 Mark MacEwen/naturepl.com; 6–7 Federico Veronesi; 8 Alex Page; 10–11 Federico Veronesi

1 THE HARDEST CHALLENGE

12–3 Paul Souders/WorldFoto; 14–5 Renaud Haution; 16–7 Will James Sooter/sharpeyesonline.com; 18–23 Federico Veronesi; 24–5 Daniel Rosengren; 26–7 Silverback; 28–31 Huw Cordey; 32–3 Mark Mohlmann; 34–5 Péter Fehérvári; 36 Ilaira Mallalieu; 37 Ramki Sreenivasan/Conservation India; 38 Patricio Robles Gil/Minden Pictures/FLPA; 39 Paul Souders/WorldFoto; 40–1 Jenny E. Ross; 42–3 Paul Nicklen/National Geographic Creative; 44–5 Brandon Cole; 46–7 Silverback; 48–9 R. L. Pitman

2 HIDE AND SEEK

50–1 Art Wolfe; 52–3 Tim Laman/naturepl.com; 54–5 George Sanker/naturepl.com; 56 Malcolm Schuyl/FLPA; 57 Donald M. Jones/Minden Pictures/FLPA; 58–9 Tom Dyring; 60–1 Pål Hermansen; 62–3 Steve Winter/National Geographic; 64–5 Suzi Eszterhas/naturepl.com; 67–9 Javier Mesa; 70–1 Emanuele Biggi/Anura.it; 72 Mark Moffett/Minden Pictures/FLPA; 74 Tim Laman/naturepl.com; 75 Jurgen Freund/naturepl.com; 76 Roman Wittig; 77 Cristina M. Gomes; 78–9 Alex Wild; 80–1 Silverback; 82 Christian Ziegler; 83 Mark Moffett/Minden Pictures/FLPA

3 NOWHERE TO HIDE

84–5 Federico Veronesi; 86–7 Daniel Rosengren; 88–9 Paul Souders/WorldFoto; 90–1 Federico Veronesi; 92 Federico Veronesi; 93 Ellen Husain; 94–5 Silverback; 96 Federico Veronesi; 97–9 Dylan Smith; 100 Ary Bassous; 101 Jonathan Jones; 102–3 Ary Bassous; 104–5 John Aitchison; 106–7 Silverback;

108 Daniel J. Cox/NaturalExposures.com; 109 Silverback; 110–1 Daniel J. Cox/NaturalExposures.com; 112–5 Will Burrard Lucas/burrard-lucas.com; 116–7 Ben Cranke; 120–3 Silverback; 124–5 Paul van Schalkwyk

4 RACE AGAINST TIME

126–7 Oliver Scholey; 128–9 Kevin Schafer/Minden Pictures/FLPA; 130–1 Tom Beldam; 132–3 Andrew Mason/FLPA; 134–5 Henk Schuurman/hscf.nl; 136–7 Silverback; 138 Kevin Flay; 139 Silverback; 140 Pete Bassett; 141 Marcelo Flores; 142–3 Mark MacEwen/naturepl.com; 144–5 Paul Souders/WorldFoto; 146–7 Oliver Scholey; 148–9 Paul Souders/WorldFoto; 150–1 Mandi Stark; 152–3 Solvin Zankl/naturepl.com; 154 Barbara Kolar/Brown Hyena Research Project; 155 Frans Lanting; 156–7 Ignacio Walker; 158–9 Silverback

5 IN THE GRIP OF THE SEASONS

160–1 Sergey Gorshkov/naturepl.com; 162–3 Silverback; 164–5 Paul Souders/WorldFoto; 166–7 Paul Nicklen/National Geographic Creative; 169 Silverback; 170–1 Paul Nicklen/National Geographic Creative; 172–3 Markus Varesvuo/naturepl.com; 174 Auden Tholfsen; 175 Mike Potts/naturepl.com; 176–7 Jonnie Hughes; 178–9 Silverback; 180–1 Sergey Gorshkov/naturepl.com; 182–3 SueForbesphoto.com; 184–5 Silverback; 186–7 Jonnie Hughes; 188–9 Silverback; 190–1 Paul Souders/WorldFoto; 192–3 Silverback; 194 Ole Jorgen Liodden/naturepl.com; 195 Erlend Haarberg/naturepl.com; 196–7 Sophie Lanfear

6 HUNGER AT SEA

198–9 Chris & Monique Fallows/naturepl.com; 200–1 Gisle Sverdrup; 202–3 Silverback; 204l Kevin Flay; 204m David Shale/naturepl.com; 204r Solvin Zankl/naturepl.com; 205l Norbert Wu/Minden Pictures/FLPA; 205m & r L. P. Madin,

WHOI; 206–7 Alex Tattersall; 208–9 Brandon Cole; 210–1 Silverback; 212–4 Gisle Sverdrup; 215 Brandon Cole; 216–7 Jamie McPherson; 219 Mark Jones/RovingTortoisePhotos; 220 Brandon Cole; 222–3 Jim Abernethy/Getty Images; 224t David Shale/naturepl.com; 224b Photo Researchers/FLPA; 225 Danté Fenolio/anotheca.com; 227t Danté Fenolio/anotheca.com; 227b, 228t Solvin Zankl/naturepl.com; 228m & b Danté Fenolio/anotheca.com; 230 Danté Fenolio/anotheca.com; 231tl David Shale/naturepl.com; 231tr Solvin Zankl/naturepl.com; 231b Kevin Flay; 232–5 Silverback

7 TALES FROM THE HUNT

236–7 Rolf Steinmann; 238–9 Richard Herrmann; 240–1 Jonnie Hughes; 242 Silverback; 243 Huw Cordey; 244–5 Mandi Stark; 246 Jonnie Hughes; 247 Silverback; 248–9 Rolf Steinmann; 250 Jesse Wilkinson; 251 Jonnie Hughes; 253 Luke Barnett; 254–5 Silverback; 256 Darren Clementson; 257 Silverback; 258 Darren Clementson; 259 Huw Cordey; 260–1 Gisle Sverdrup; 262–3 Richard Herrmann; 265 Silverback; 266–7 Hans Rack; 268–9 Adrian Seymour; 270–1 Håvard Festø; 272 Sophie Lanfear; 273 Silverback; 274–7 Huw Cordey; 278–80 Gisle Sverdrup; 281 Ignacio Walker; 282–3 Silverback; 284–7 Jonnie Hughes; 288–91 Ellen Husain; 292 Silverback; 293 Ellen Husain; 294–5 Oliver Scholey; 296–9 Sophie Lanfear; 300–1 Silverback; 302–3 Adrian Seymour

Endpapers: front Federico Veronesi; back Brandon Cole

Research permits: (blue whale USA) National Marine Fisheries Service 16111; SEMARNAT (blue whale Mexico) 01577; (humpback Australia) Commonwealth Marine Reserves 2013/06844; Department of Parks and Wildlife FA 000114